A Road Map to the Development of European SME Networks

Agostino Villa • Dario Antonelli
Editors

A Road Map to the Development of European SME Networks

Towards Collaborative Innovation

Springer

Professor Agostino Villa
Professor Dario Antonelli

Department of Production Systems
and Business Economics (DISPEA)
Politecnico di Torino
Corso Duca degli Abruzzi 24
10129 Torino
Italy

ISBN 978-1-84996-789-1 e-ISBN 978-1-84800-342-2

DOI 10.1007/978-1-84800-342-2

British Library Cataloguing in Publication Data
A road map to the development of European SME networks :
towards collaborative innovation
1. Business networks - European Union countries
2. Government aid to small businesses - European Union
countries
I. Villa, A. II. Antonelli, Dario
338.8'7

Cover design: eStudio Calamar S.L., Girona, Spain

Printed on acid-free paper

9 8 7 6 5 4 3 2 1

springer.com

Preface

Recent studies supported by the European Commission have reported that half of the small and mid-sized enterprises (SMEs) cooperate with other SMEs, with a traditionally large diffusion of collaborative agreements in countries such as Italy, Denmark, Norway, Finland, an increasing diffusion of similar arrangements also in France, Germany and in some new EU countries of Eastern Europe, such as Hungary (see the report "SMEs and Cooperation", published by the Observatory of European SMEs in 2003). A frequent motivation for SME cooperation is to have better access to larger markets, a cheaper supply of materials and components, but also a reduction of costs for various services and a smooth and assured meeting of client demands. The same report shows that the size of the enterprise can influence the type and scope of the cooperation: the main goal of mid-scale enterprises is to have long-term agreements, whilst small-scale enterprises want to have short-term benefits, mainly through market-oriented coordination. On the other hand, some obstacles exist: foremost, each SME wants to maintain its independence, and this barrier is an evident cultural motivation, difficult to eradicate. However, in spite of some obstacles, the idea of SME aggregations into clusters and networks, and an increasing trust in these groupings, is growing in all European countries. More research recently published by the Directorate General Enterprise and Industry of the European Commission gives a clear insight in the dissemination of regional/local clusters, which are being recognized as important tools for industrial development (see "Innovation Clusters: A Statistical Analysis and Overview of Current Policy Support", 2007, at http://ec.eurpa.eu/enterprise/newsroom/cf). A SME cluster can now be clearly identified as an aggregation of co-located producers and service providers, often including educational and research institutions as well as financial services and also public institutions. The 2007 report presents a statistical picture of clusters in Europe, and shows how they are becoming real drivers of growth, mainly to enhance not only cooperation among firms but also to strengthen links with the knowledge infrastructure to push innovation. However both reports stress the need for a more precise conceptualisation of what constitutes a SME cluster and which main phenomena characterize its evolution, i.e. their desired cultivation but also their collapse. Research and study efforts to accurately identify the industrial body "SME cluster and network", together with analyses of the real causes of weakness or strength, appear to be mandatory not only for supporting both the design and implementation of new cluster-oriented policies, but also for giving suggestions and criteria for the constitution of new SME aggregations, and for driving an effective organization of SME interaction and cooperation.

The EU-funded Coordination Action CODESNET (COllaborative DEmand and Supply NETworks), developed during the 6th Framework Programme, has been developed in order to promote a dissemination of these ideas, with the contributions of 22 industrial enterprises and academic institutions of 11 European countries. Theoretical considerations and a wide collection of SME networks from the partners offered a promising ground on which a new methodology for analysing these networked industrial bodies, such as evaluating their performance, can be developed. The collaborative actions done during the project generated a preliminary version of a method for comparing different SME networks and identifying the best characteristics of each. Potential applications of this method have been validated by implementing it through a specific web portal (see the address: http://www.codesnet.polito.it). This book aims to spread the project ideas and tools of CODESNET, putting them at the disposal of any industrial analyst interested in promoting these very important bodies: the SMEs and their cooperation opportunities. To this extent the book, whose contents have been planned during the final period of the project development, includes contributions from several project partners. Starting with a summary of the main motivations, which suggests the promotion of research and studies on SME aggregations, it presents a view of SME clusters and networks in a number of European countries, together with suggestions and hints towards a better diffusion of these collaborative industrial bodies. The CODESNET models and concepts for the networks performance evaluation are then addressed, oriented in a comprehensible form for industrial end users.

Acknowledgments

The book co-authors are all greatly indebted to Dr. Florent Frederix, Commission Officer of the CODESNET project from the European Commission, who gave invaluable support to the project organization and development. All co-authors also wish to thank Prof. Paul Valckenaers, K.U. Leuven PMA, Dr. Anna Psoinos, European Central Bank, and Dr. Herbert Heinzel, H_2O BTC, for their helpful comments and suggestions.

Contents

Chapter 1
Analysing Collaborative Demand and Supply Networks in a Global Economy

Abstract In several European countries, interest in small and mid-sized enterprises (SMEs), their importance in national industrial and economic systems as well as their effective impact on both local and Europe-wide markets, has been growing for several years. More recently, the globalization of markets has had dramatic effects on many small enterprises, mainly on those industrial bodies not characterized by either niche-products or by high-level technologies. In general, any type of small and mid-sized enterprise has suffered from low-price products flooding the market from emerging industrial powers. Since the end of the last century, the European Commission has promoted studies on the situation of SMEs with the goal of investigating how small industrial bodies, which are widely distributed in Europe, could reinforce their standing through aggregations, consortia agreements, collaborative networking and so on. This chapter describes the main characteristics as well as the most significant motivations of SME evolution from individual companies, too weak in an international highly competitive market, into clusters and networks. Analysis of the SME need for clustering and networking is given in Sect. 1.1. The analysis is presented not only in terms of industrial needs, due to the recent market enlargement, but also in terms of local and regional desire to make the labour markets, and not only the product markets, a better representation of local abilities and specializations. Examples of SME aggregation into networks are presented in Sect. 1.2 to illustrate how the SME evolution is continuing to this day. This evolution, and the main organizational problems which it is now forced to approach, is a key motivation of this research, as discussed in Sect. 1.3.

A. Villa, D. Antonelli, *A Road Map to the Development of European SME Networks*,

1.1 SME Networks: a European Way to Collaborative Innovation

A. Villa and D. Antonelli – Politecnico di Torino

At the beginning of the new millennium, industrial systems of developed countries all over the world are going to consolidate in a number of different organizational structures, which in turn also generate different dynamics. The characteristics of these different dynamics are illustrated in four main types of industrial systems, which span the globe. First, there is the "Pacific rim" area, birthplace of the "imperial" made-in-the-USA system, mainly based on very large enterprises and generous private capitals. The second example is the OPEC system control of oil, whose historic core lies in a number of Middle East countries with ramifications in Russia, north Africa and Central and South America, everywhere characterized by a strong central government, single production and a large socioeconomic disequilibrium. Third, beside these systems are the emerging countries of "state capitalism" led by the Chinese, with a tumultuous development of enterprises driven by rigid Taylorism in social environments, based on the so-called real socialism regimes.

The fourth industrial system example is the European Union, which presents a "federal" industrial environment in which some large enterprises live together as a large number of small and mid-sized enterprises (SMEs), in a social frame where overcoming barriers and borders, plus a common currency, are enabling interactions of different cultures, and pushing the EU towards more and more cooperation.

Indeed, the presence of a large percentage of SMEs inside the industrial fabric (for a number of enterprises, a percentage of SMEs greater than 90%, and for a number of employees, a percentage greater than 70%) drove the European Commission to promote studies and research programs on SMEs and the SME clusters phenomenon, in order to assess their strengths and weaknesses, and to develop new instruments or strategies to promote their growth.

Among these initiatives, the EU-funded CODESNET project (on whose results this book is based) and the recent report of the European Cluster Observatory (European Cluster Observatory Report 2007), give a good picture of the industrial clusters in Europe.

The following is a definition to which all the following chapters will refer. According to the European Cluster Observatory Report, "clusters are defined by the co-location of producers, services providers, educational and research institutions, financial institutions and other private and government institutions related through linkages of different types." Indeed, industrial clusters could show a significant difference from one another: SME networks differ from clusters organized around a leading firm, and also from high-tech SME groups developed around a university. But every industrial cluster is characterized by a wide presence of SMEs.

It could appear as a paradox that globalization is strengthening the importance of clusters in every European country, but this expanding situation can have clear justifications. All enterprises find it promising to locate their sites and core activities in areas characterized by a good business environment. The appeal of a business environment increases if the area includes other companies, either complementary or similar, and also "real services" together with a specifically trained local labour market. Such features make an area attractive, facilitate multiple allocations and force a regional specialization: all conditions which make the enterprises in the area more competitive and more efficient in a worldwide market, even if they are small in size (European Commission 2002, 2003).

Due to cluster development, several European regions in the recent past have gained a real competitive advantage in specialized activities and, as a consequence, have been able to attract high-quality people and new investments, apply new technologies and increase market shares. This fact highlights the importance of industrial clusters and their association with a higher level of innovation and prosperity. As reported in the European Cluster Observatory Report (2007), between 30 and 40% of all employment is located in enterprises concentrated in industrial clusters. Approximately 38% of all European employees work in enterprises that are components of clusters, and more than 20% of the employees belong to regions that are more than twice as specialized in a particular cluster category than an average region.

An important feature of successful clusters all over Europe is the presence of a number of collaborative actions among the enterprises belonging to them. Collaboration indeed implies that several enterprises operate together in concert in order to execute critical business processes by sharing responsibility, accountability and production quality. The introduction or, more simply, the identification of collaborative procedures in the relationships among firms is widely acknowledged to have a positive effect on firm efficiency, quality and profitability. The evolution of the collaboration paradigm started from the collaborative supply chain in which the client firms spread information about their production programs globally over the chain and not only to their direct suppliers, allowing a better logistic organization of the supply chain as a whole. Present evolution extends the concept of collaboration to the entire cluster or to part of it, by creating a collaborative demand and supply network in which there is a widespread sharing of information among all the clients and all the suppliers of the network.

In addition to these data, which give an overview of the clusters' situation at the end of 2007, industrial clusters are not static, and new ones may emerge over time in areas that previously seemed to be unattractive. The cluster dynamics must be better understood in order to define regional development strategies and any type of industrial policy that aims at facilitating developments and structural industrial changes.

The above considerations, and mainly the need for better knowledge of the present status and future dynamics of industrial clusters, opens three types of questions for researchers and managers, as suggested in the book by Thomas Brenner (Brenner 2004):

1. Why do local industrial clusters exist?
2. When and where do industrial clusters emerge?
3. How will industrial clusters develop in the near future?

The third question has been largely studied and investigated in literature, starting from the works of Italian researchers on a special type of cluster called industrial districts, agglomerations of SMEs widely diffused in several Italian regions. As analysed in the following chapters, some authors since the 1980s (Becattini 1987, 1989; Brusco 1982) have studied the motivations and success of this special industrial body aggregation. In the many of the references cited in this book, various definitions and characterizations of the industrial clusters have been reported, each one trying to emphasize one or more specific features. This variety of definitions has been reinforced by the fact that when analysing SME aggregations, several types of industrial clusters are described (Bergman and Feser 1999) in the evolution of European industries, also named different ways depending on the country: industrial districts in Italy, network of competence in Germany, poles of competitiveness in France. All these names represent a European way of aggregating SMEs to compete in a worldwide business market, leaving at the same time autonomy to each SME (Albino and Kühtz 2004; Picard and Toulemonde 2003; Rosenfeld 1995). Flourishing throughout Europe, even if firstly only documented in Italy (Paniccia 1999), industrial districts have attracted much attention in the academic literature. Therefore, for the necessity of defining these industrial bodies and for the sake of classification, academics first directed their interest to the present status and expected development of SME aggregations, by trying to identify different types according to their features and the kind of interactions among enterprises.

As the second question is concerned, a large body of literature presented evidence of successful industrial districts in Italy and elsewhere, and of their ability to sustain national competitiveness in several countries in the past decade. Other research dealt with the comparison, from the economic perspective, of the performances of firms with respect to their belonging to a district as in the work by Fabiani and Pellegrini (Fabiani and Pellegrini 1998). Other authors (Molina-Morales 2001) suggested the hypothesis of good profitability for SMEs belonging to districts in terms of ROE, ROI, etc. Unfortunately analysis carried on different sets of clusters gave an opposite set of results (Love and Roper 2001), namely enterprises belonging to clusters in UK and Germany have comparable return on investments (ROI) for their innovation activities with the enterprises which compete on the market on a stand alone basis. That is why the attention of some authors shifted from economic analysis to the consideration of the district performances from a social viewpoint, in order to better understand the influence that the district has on the population in which it is set (Nohria and Eccles 1992; Granovetter 1985).

The questions of why local industrial clusters exist and which are the best organizational conditions that could promote cluster development, have not often been addressed in the literature. Some works approached this problem by analysing several real clusters and trying to apply statistical tools to identify certain conditions (Paniccia 2002; Antoldi 2006).

In particular, a few studies approached district analysis from the crucial viewpoint of the organizational arrangement, and suggested the opportunity of a comprehensive review of different roles that governance should have in an industrial district, by examining the relationships between the district's features and the organizational performances.

If one would try to gather only the main concepts treated in the literature on SME aggregations, a short list of main competitive assets of industry clusters could be summarized as follows:

- Within the district there is a division of labour among firms, which promotes high levels of flexibility and productivity.
- High degree of specialization in one or few complementary industries.
- Horizontal competition and vertical cooperation: the spatial concentration and strong complementarities among different units turn competition into a connective force among agents.
- A distinctive milieu that includes the local institutional infrastructure.
- Common marketing strategies.

But, if one would like to justify the real motivations of such a list of competitive assets, no sufficiently assessed formulations or robust demonstrations of the issues could be found.

From this lack of practical tools for both industrial managers and policy makers the CODESNET project was created. The scope has also been specified by the following consideration: from the analysis of a wide set of industrial clusters composed of SMEs, the main result useful for an industrial analyst is not to compute some statistics, but to generate a formal model which could describe both the general characters of a SME cluster and its specific features, and to allow the identification of both the principal organizational/technological parameters and the most significant performance indicators.

But, first, let us consider another question that many analysers do not take into account, even if it seems crucial to understanding any evolution: what is the industrial history of European industrial clusters?

1.2 Industrial Districts: Historical Tools for Complex Systems

V. Marchis – Politecnico di Torino

Industrial district is a concept that has been widely investigated from the point of view of business organizations, of urban planning, of economic history and of industrial archaeology. Recently, territorial dynamics of industrialization was the theme of a conference held in Helsinki in January 2006 and organized by Jean-Claude

Daumas, Pierre Lamard and Laurent Tissot (http://hsozkult.geschichte.hu-berlin.de/termine/id=2336). H-Soz-u-Kult – Kommunikation und Fachinformation für die Geschichtswissenschaften (Communication and Information Services for Historians; http://hsozkult.geschichte.hu-berlin.de/index.asp?pn=about) is a network that offers a vast archive of studies for historians interested in history of science and technology on the Web, especially from a social and contextual point of view.

Nevertheless, the history of industrial districts and the territorial dimension of productive organizations remains a topic strangely neglected. This is probably due to the difficulties emerging from an interdisciplinary approach, where many disciplines require a historical methodology that often is forgotten by technicians. On the basis of the historical approach, the industrial culture, even in the present time, remains a tacit culture. We focus here on some aspects of industrial districts from their dynamic (historical) perspective and emphasize which are the documents to be used both by historians and technicians to investigate these bodies. The consciousness of the *present state* of a social system resides in the past, and the more recent time, unfortunately is often neglected in a critical analysis. For these reasons the industrial district history becomes mandatory, first of all in the engineering context, where the narrative dimension of facts is often considered as accessorial and marginal. But history is not only a list of *facts*, but also their *investigation* and chiefly in their *telling*.

The *scientific* storyteller point of view requires a severe choice of facts and figures, and they have to be selected and supported by documents that involve many categories of data records: they are the basis of our memory management.

From the early eighteenth century, the information gatherers started to report on societal changes due to the Industrial Revolution and between them Daniel Defoe was certainly the most famous. His "A Tour Thro' the Whole Island of Great Britain", first published 1724–1726, and described as "the liveliest introduction … to Britain in the early eighteenth century" (Defoe 1724), contains so many notes about the birth of "industrial districts" that this book can really be considered as a first map of their impact on territory planning. The industrial district of West Riding, in Stephen Caunce's paper "Revealing A New Northern England: Crossing the Rubicon with Daniel Defoe" (Caunce 2007) emerges as the most dynamic in Europe despite its apparently desolate Pennine surroundings, and is portrayed by Defoe in a particularly striking manner, where he argues that the region was very misunderstood. Though this section of the "Tour" is often quoted, few people have analysed it in depth, or linked it to later developments, including the classic Industrial Revolution. This paper sets out to show that, while Defoe's topography and factual description can only be used with care, as a study of a remarkable nascent business culture it is of vital importance, and emerges on this level full of insights that can be obtained nowhere else. Within a holistic account, that no other source can rival, Defoe brings graphically to life the way industry was embedded in rural life, rather than destroying it: "Defoe described the Halifax area in the 1720s: clothiers held some land in the woollen districts, and there was thus some diversification in sources of family livelihood, but farming activity was insufficient to feed the area, and the result was to make the textile West Riding a major target for food suppliers from the east for grain and the north for meat." This information helps us to understand, at the roots level, how the changes developed during early times and gives important clues for investigating more recent events. Texts, and in particular the classical literature, give little attention to the industrial facts at that time and therefore they remain in complete oblivion. Reports about the industrial *state of the art* and their technological or environmental scenarios are very rare, and also today the specific literature presents only quantitative figures, and judges that other narrative forms are poor in scientific insight.

Alan Everitt wrote an essay entitled "Country, County and Town: Patterns of Regional Evolution in England" (Everitt 1979) and again it paradoxically is easier to understand these earlier industrial changes than more recent ones, where the "context noise" and the lack of documentation leaves a sort of Alzheimer pathology to this knowledge area.

Again, if the primary literature in this field is poor with regards to British development, it is more and more scant with reference to the Italian area, where nevertheless the studies conducted by Luciana Lazzeretti emerge (Lazzeretti 2002).

Together with Vicenza and Valenza Po, Arezzo is one of the Italian jewellery industrial districts (Fig. 1.1). As Lazzeretti described in her paper (Lazzeretti 2003), for investigating this district the HEDRON (Historical, Economic, DemogRaphic, ecOlogical, ecoNomic) Methodology was adopted along five multidisciplinary phases:

- *Phase 1.* Historic and economic analysis. The analyst could approach the main evolutionary stages of the event under study, through the analysis of the his-

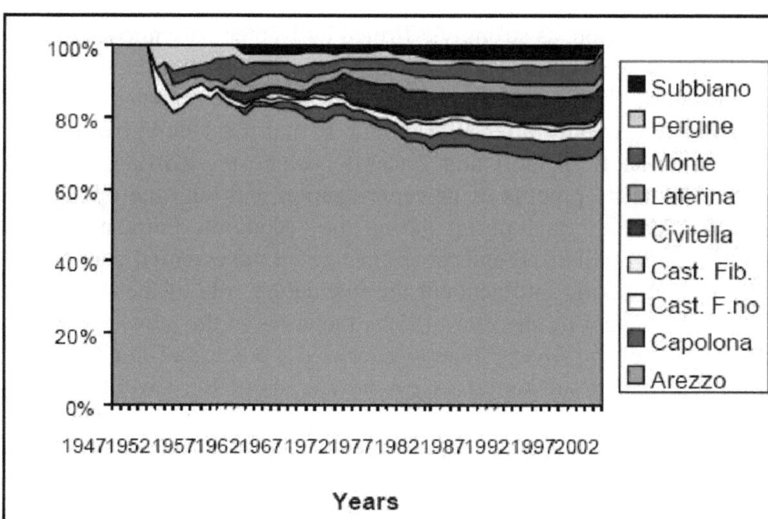

Fig. 1.1 Evolution of jewellery firms as to their localization, from 1947 to 2001 (adapted from Lazzeretti 2002)

torical and economic literature, such as to identify a valid date from which to start the investigation.

- *Phase 2.* Economic and industrial analysis. The analyst can provide an exact individuation of productive *filières* in the various periods, and identify the populations to be monitored, and the geographical area to be taken into account.
- *Phase 3.* Demographic and industrial analysis. To be developed by use of data relative to births/deaths, entries/exits, and other economic variables under evaluation, in a first stage of statistical processing. From the statistical results, if any ecological theory can find confirmation, then the analyst could go on to the next phase, otherwise only descriptive statistics should be used in order to support qualitative studies.
- *Phase 4.* Ecological analysis. To be developed when a statistical model can be defined and used in order to test the ecological theories taken into account such as, subsequently, to proceed to the ecological interpretation of the phenomenon under study.
- *Phase 5.* Economic and industrial analysis. To re-interpret the demographic and ecological results from an economic and industrial profile, in order to supply useful information about the evolution of sectors and local communities of populations.

Even if the above reported study asserts that "at this first stage of analysis, we explored the applicability of the ecology-based approach rather than going deeper into the qualitative understanding of the event under study, and so we privileged the 'robustness test' (phases 3 and 4) over the demographic and ecological aspects, leaving out the historical and economic ones", the analysis of the events remains a basic step to always be considered in order to avoid a black-box approach. *Inside the Black Box* always remains as a mandatory paradigm in any economical-social-technical system investigation, as strongly affirmed by Nathan Rosenberg (1983).

During the Fascism era, within the dominance of agriculture over the Italian economy, the birth of the first Aretinian jewellery firm must be recorded: Gori and Zucchi (1926). But it was only after World War II that the UnoAErre industry (after the transformation from Gori and Zucchi) became the starting point for a modern district. With the process of de-centralization and with the opening of manufacturers of goldsmith machineries during the "economic miracle" several artisan laboratories and small-sized enterprises generated the essential core for the industrial "sector" that, taking into account the mandatory role of the trademark, enters the system phase. During the early 1970s, the crisis in the jewellery sector, and the positive action of the silver processing, caused a process of fragmentation and emphasized the role of subcontractors. A new crisis of the jewellery sector (1979) forced a restructuring and the passage to the "district". In the 1980s the Arezzo area changed into a system of small firms and started a process of diversification with growth of exports. Again a new crisis of the internal market (1992) caused an enhancement of structural and occupational development, which strengthened the identity of the industrial district.

Another history can be reported for the Valenza Po district (http://www.italtrade.com/focus/valenza.htm), in the province of Alessandria (Italy), where around 40% of the working population are employed in businesses operating in the gold and jewellery sector. The origin of this site vocation dates back to 1840 when Vincenzo Morosetti began working as a goldsmith. But it was his apprentice Vincenzo Melchiorre, who learned and perfected the trade in Turin and Paris, who started a high-quality jewellery production in Valenza, where in 1850 three companies were producing gold jewellery. In 1913 they had become 44, with 517 employees, out of a population of 5,000. In 1945 the small companies grew up to 300 and the Valenza Goldsmiths' Association was created for a common production and marketing policy. Today, it offers a wide range of services and represents over 700 companies in the Valenza area where each year 30 tonnes of gold and 80% of the precious stones imported into Italy are elaborated.

A comparison between the Arezzo and the Valenza Po industrial districts, both operating in the same field but showing different peculiarities, also from the historical point of view, emphasizes an aspect often forgotten by the most recent analysis. The *point of origin* of a production activity capable of generating an industrial district must be linked to the cultural ground, which is a blend of several components: previous industry or handicraft, territorial characteristics, social environment, and critical events.

In a recent book, Dario Gaggio through a comparison with the jewellery district of Providence, Rhode Island, shows that these Italian towns were not unique in the ways they navigated the challenges posed by the embeddedness of economic action in the fabric of social life.

According to Gaggio's considerations, called either industrial districts or clusters or area systems, these local networks represent not only an extremely vital component of the Italian economy, but they also offer researchers the opportunity to reconsider several important theoretical questions (Gaggio 2007). First, the many successes of SMEs belie the predictions of some intellectual groups, who viewed the large integrated corporation as the inevitable outcome of the economic evolution. Second, the networks of small producers in a localized area were able to share the benefits of external economies and solidarity ties without neglecting the efficiencies linked to competition; therefore, industrial districts are first of all historically rooted communities built on mutual trust. In the Italian context, industrial districts are now the demonstration of what Arnaldo Bagnasco calls the "Third Italy", the areas of the central and northern parts of the country located neither in the north-western "industrial triangle" (Turin-Milan-Genoa) nor in the underdeveloped regions. Then, since these pioneering studies, industrial districts have come to constitute the most powerful counterexample to the once-dominant expectation that small-scale firms would not withstand the competitive pressures of modern capitalism. Jewellery production in Providence was embedded in networks of local and extra-local social relations as in Italy. In the course of the twentieth century, these connections became increasingly ethnicized (due to the link between jewellery production and the communities generated by immigration) and gendered (due to extensive use of housework). After the Second World War, the area of Providence de-industrialized and shed much of its population. As a consequence, jewellery production became the city's largest manufacturing industry.

Many studies have recently been performed around the industrial districts but it is difficult to identify a general rule for identifying the most important features that can be assumed as characteristic for each situation. Every study appears as a single and unique case; the study "Guild's Lake Industrial District. The Process of Change over Time" (http://www.historycooperative.org/journals/ohq/107.1/dibling.html) remains a local history and any comparison with other similar studies becomes very difficult (Fig. 1.2).

Returning to Italy, a few words can be spent around the Suzzara industrial district.

The Region Lombardia is subdivided into several districts: among them District #21 has an area of 280 km² and is located between the rivers Po and Secchia, belonging to seven municipalities: Suzzara, Gonzaga, Moglia, Motteggiana, San Benedetto Po, Pegognaga and Borgoforte. The typology of this district is devoted to the metal carpentry and to the agriculture machines, ranged by over 260 units and about 3,000 employees. The mechanical industry grew on an agriculture and zootechnic tradition that has been characteristic of the area for a long time, notwithstanding the decrease of activities in this sector that converted their funds into new technologies. The typology of the small industries of the Suzzara district ranges also in parts and mechanisms for machines and plants, and this gives the district great autonomy and flexibility.

In the mid-nineteenth century the Officina Casali was founded, which had a great influence on the industrialization of the local agriculture: before the First World War it became the most important industry for the production of harvesting

Fig. 1.2 Guild's Lake industrial district in 1955

machines in Italy. This factory, transformed into the MAIS (Meccanica Agricola Industriale Suzzarese) during the 1920s changed the morphology of the local territory drastically. After World War II the mechanical industries settled in the Mantua province for the conversion of the factories, and the FIAT entered to control the district. When during the workers crisis of 1955 FIAT decided to reduce the number of workers, new small artisan enterprises grew up and formed the new model for the FIAT *indotto* (the Italian word to define the cluster of small and mid-sized enterprises that produce parts of FIAT vehicles), which has since characterized the Suzzara district (Fig. 1.3). In September 2001 the district suffered a critical phase owing to the sanitary emergency due to BSE, to the very bad meteorological and climate conditions, to waiting for the political elections, and to the foreseen investment policies. The extremely dynamic social context of this district, especially during the recession times, has been capable of taking up the workmen and dumping the social conflicts, and also in actuating new diversification processes as in hosiery, in wood furniture and again in metallic carpentry.

An ecological analysis of the evolution of an industrial district, rolling over its birth, growing, ageing and death, must lead to a basic consideration: any industrial system has to be considered as a living system, whose conception may be casual, but its surviving capabilities have to be sustained by a concurrence of several

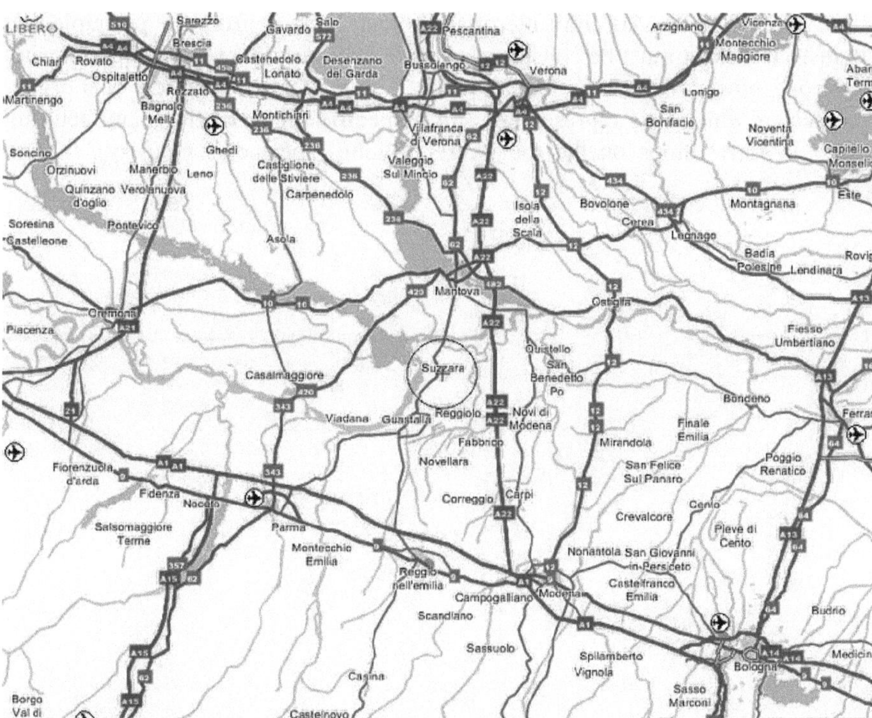

Fig. 1.3 Location of the Suzzara district near Garda lake

causes, obviously intentional and well-organized. Why do historians not examine the case of those industries whose success failed? Probably in the particular example of the examined Italian "gold industrial districts", due to the nature of the products requiring no large efforts in the territorial infrastructures, the geography of the land did not play an important role. But this is not a general rule. The geographical position of a successful industrial district is a central point to be examined, because freight transport lines and the energy network are the connective fabric on which productive activity develops itself.

By considering the strong acceleration of the dynamic processes involving the industrial districts, even in the presence of the "information and knowledge revolution" that changes today's social paradigms, memory of the more recent past becomes of prime importance for any future planning.

Since at the heart of the industrial development lay cultural and social reasons, which link populations to the territory, avoiding the migration from the periphery to the large town agglomerations, beside the development of infrastructures for both freight and information transfer, policies for promoting local social network have been started in Italy (Fig. 1.4). In 2002 the Italian artist Michelangelo Pistoletto with his "Manifesto of Art and Enterprise", started a collaboration between The Club of Industrial Districts and Cittadellarte – Fondazione Pistoletto, in order to join the art and business worlds and to enhance the cultural qualities of Italian products. The Club of Industrial Districts replied to Cittadellarte's proposal, giving rise in 2003 to the Art and Enterprise Project. Starting from the principle that an industrial district must be conscious of its own cultural power, and that the arts have strong power in their prophetic action, a tour of seven districts was organized, each of which was representative of a specific sector of the manufacturing industry in Italy. Stops on the tour were Belluno (spectacles industry), Matera

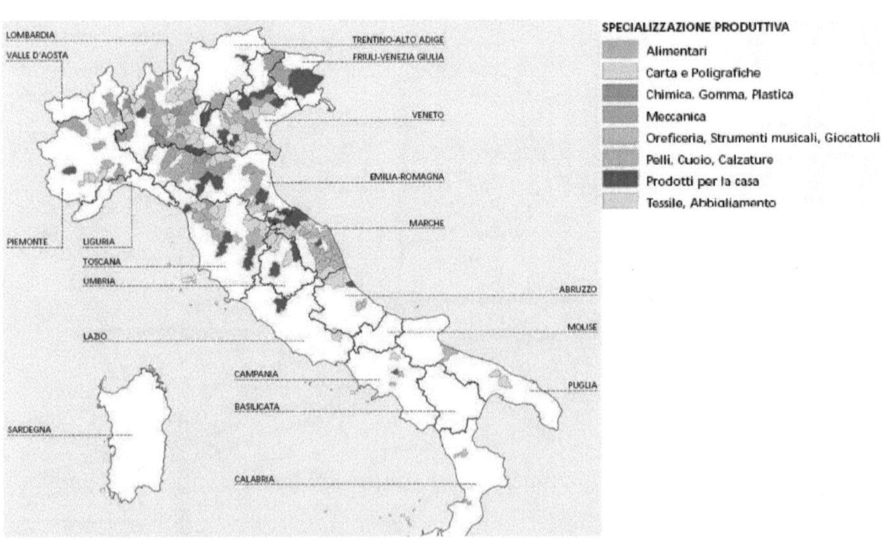

Fig. 1.4 Industrial districts in Italy

(furniture), Nocera Inferiore (agro-industry), Valenza (gold working), Cusio Valsesia (household goods) and Fermano Maceratese (shoe manufacturing).

Even if these de-structured considerations appear chaotic and without a precise order, the aim of this work is to revaluate the narrative dimension of the industrial memory, because the quantitative analysis, in spite of any deterministic approach, is not capable of describing with synthetic means what happens in an industrial district.

As a conclusion, we present an approach to the problem that probably will appear understandable to engineers only. The solution of a dynamic system becomes impossible when initial conditions are unknown, and these *conditions* from a mathematical point of view, look more at the past as a higher degree of the differential equation model of the system itself. Unfortunately the industrial culture operates as if it were an algebraic structure, without important links to the past, and therefore demonstrates a low sensibility in preserving its own memory. Only a critical policy of industrial archives appears as a way for finding more realistic interpretative models, absolutely necessary for planning the future.

In the analysis of a complex system, as is the evolution of industrial districts, the more information available, the greater the possibility of understanding the present. In the information revolution of the present we must remember that not only the figures (data, tables, inquiries, etc.) have to be taken into account but also the pictorial and visual memory, which involves drawings, maps, photos, and movies.

1.3 Motivations for a New Analysis Methodology

A. Villa and D. Antonelli – Politecnico di Torino, Torino (I)

The historical picture above (Fig. 1.5) has shown that although active since the Industrial Revolution, only in the last twenty years has it became apparent that SME clusters add competitive assets and are stable organizations for small enterprises in order to compete in the worldwide business market.

This is the reason why the literature on industrial SME aggregations presents has a large number of contributions – books, papers and so on – as outlined in the references of this book. However, despite the crowded set of studies and reports, investments in analysis and research on these industrial aggregations are further advanced, because they present some causes of potential crisis, not only conditions for real development.

This intrinsic ambiguity still remains to be clarified, owing to several motivations, all to be referred to the intrinsic complexity of a SME cluster.

On one hand, SME clusters are really complex dynamic systems, because they involve a number of enterprises, each one with a proper autonomous management and proper goals and constraints, but all connected together by some agreements.

Fig. 1.5 Kansas City industrial district during the flood (July 13, 1951). The water surface clearly shows industrial plants and territorial infrastructures

In theoretical terms, a SME cluster could be viewed as a network of individual "agents", that are individual decision-makers who try to optimize their own objective and reach their own target (Haddadi 1995). The network connections induce on each decision-maker some limitations on its own autonomy in adopting decisions, plans and control strategies of its own firm. Then, the SME cluster should be a system organized in such a way to be able to be conditional upon each individual decision-maker: for each SME, this implies to define and accept agreements in order to belong to the cluster itself.

On the other hand, a SME cluster can also be viewed as a complex set of local objectives that are management targets planning the production and service goals to be reached. But, for a SME, to be included in a cluster means to realize that a local objective must be made coherent with the ones of other SMEs, at least as coherent as possible. In theoretical terms, this situation could be described by considering that the individual objective of each decision-maker in the abovementioned set must be modified in order to not conflict with the objectives of other decision-makers (Villa et al. 2005). In practice, this shows the need for establishing a collaborative operating system for all SMEs included in the cluster. But both to state and to verify existence of an effective collaborative behaviour of all SMEs is an open problem. Indeed, the concept of *collaboration* still remains not so well-defined.

Another cause of complexity of a SME cluster comes from the nature of a cluster as an industrial complex body, that is, a set of SMEs, interacting with each other but mainly with strong interactions with their own markets. And, as the set of clusters presentations contained in the CODESNET website can show, the number of different markets with which the clustered set of SMEs is connected is as large as the production and service variety of the cluster itself. All these interactions could be potential causes of conflicts among the internal SMEs. In addition, several interactions of some internal SMEs together are market transactions themselves (as from a supplier of product components to a final product producer). All these market transactions are potential causes of instability and crisis for the cluster: then, they must be made "coherent" to each other. This is the conceptual idea of "interoperability", whose translation into practice is pushing research on information management networks able to make each SME management system able to clearly and unambiguously interact with the others (INTEROP 2007).

All these complexities justify a certain lack of methods and procedures to be applied in a practical but accurate analysis of the SME cluster performance, especially if dedicated either to support an efficient/effective/convenient management of the cluster or to define a cluster innovation program. This present lack of practical tools for SME cluster performance evaluation is one of the main motivations of the new analysis methodology developed in the CODESNET project, and here presented in its developed form.

But, in order to approach such a complex system for purpose of analysing its performance in such a way to be able to adopt effective management strategies, one must not use a unique viewpoint, but have a multi-faced description of a SME cluster. Indeed, the existing literature supplies at least three complementary viewpoints (Antoldi 2006):

- An *economic approach*, which considers an industrial cluster as a peculiar organization of the production activities, characterized by the existence of agglomeration economies both external to individual SMEs, and internal to the local set of SMEs.
- A *sociological approach*, which is mainly interested in the social structure on which the production structure is based, so that main aspects analysed are the so-called social capital and the knowledge generation and transfer in the SMEs network.
- An *industrial approach*, which aims to analyse the production and organization aspects of both the SMEs and their network.

These three viewpoints are usually adopted by different research environments, and have first found a large number of applications in analysing the Italian industrial districts, then have been extended to the analysis of any other SME agglomeration in Europe.

So, when trying to motivate a new performance analysis methodology, the problem to be first considered is which type of contributions could be gained from already existing approaches, and which further development should be searched for.

If one moves according to the *economic approach*, two economic concepts have to be taken into account: that of *aggregations economy*, referred to as a view of spatial distribution of activities and SMEs on a geographical area/region, and that of *transactions (or external) economy*, referred to as the potential advantages/disadvantages arising from the labour activities subdivision and specialization among the various SMEs.

A SME cluster, and typically an Italian industrial district, is a singular agglomeration of economic activities in a geographical region caused by either historical and cultural factors, as a local tradition concerning the production of a specific artefact, by natural and logistic factors, as a concentration of transportation/traffic links and nodes, or by a combination of several factors. In this case, this concentration of activities and industrial sites promotes an increase of agglomerations economies, which consist of advantages and benefits for the individual SMEs induced by their respective location. Often proximity advantages in turn foster the increase of the cluster dimension and its robustness, efficiency and competitiveness (Asheim 1994). Among this type of enterprise agglomerations, the local industrial clusters present a specific character: SMEs included inside are not only specialized in one industrial sector, but also linked by "vertical" and "horizontal relations", which means respectively supplier-to-client relations (as in a typical supply chain) and relations among potential competitors of a same product/component (as in a group of parallel producers). From this viewpoint, the Italian industrial districts appear to be production systems where the geographical factor is surely among the most important. This character differentiates them from large regional clusters, as discussed in the European documents mentioned above (and in http://europa.eu.int/comm/enterprise), whose lean structure only depends on the fact that all enterprises inside belong to the same sector or chain.

The evaluation of the cluster performance in the case of a supply chain has received considerable attention and is supported by some practical tools: among them, one can refer either to the *balanced scorecard procedure* (Bullinger et al. 2002) or to the *Supply Chain Council approach* (http://www.supply-chain.org). On the contrary, the same issue in the case of a cluster with a balanced composition of SMEs is still an open problem.

Another characteristic of SME clusters/districts arising from the economic approach results can be recognized by considering the production activities organization driven by a Taylor-based labour operations subdivision and allocations to internal SMEs. Consequent interactions among SMEs, now specialized firms in different labour operations, can be arranged either in an *internal market* or in an *internal hierarchy* (Villa 2003). When the whole production process can be decomposed into separated phases/operations and these are allocated into individual firms (the so-called autonomous agents in the large-scale systems theory (Silijak 2007)), then the enterprises within the aggregation interact together through business transactions. The elementary form of *coordination* that follows, is just a market: now each agent aims to maximize its own profit leaving the common search of all agents for an equilibrium between demands and supplies the task of harmonizing individual interests into a common frame. However sometimes markets do not behave perfectly, owing to presence of conflicting interests of enterprises characterized by different (unbalanced) strength. In these situations, some strong self-interest can generate difficulties in the marketing operations, which are pushed far from an accepted equilibrium and subjected to large transaction costs. These could be causes that force some agents to prefer a *hierarchy* instead of a free market, thus giving rise to a cluster organization based on vertical links driven by the largest enterprise: the leader-driven supply chain. Industrial districts indeed represent an organizational form in between a market and a hierarchy: each internal SME can fruitfully participate in the internal transactions but a commonly accepted agreement dedicated to assure a real cooperation must exist, to reduce as much as possible self-interested behaviour. Due to their intermediate nature, an evaluation of their performance as well as a robust support to define their coordination strategies is often an open problem, thus giving a second strong motivation to the approach proposed in this work.

According to the second analysis approach mentioned above, namely the sociological approach, the industrial clusters appear to be very complex social structures able to give rise to collaborative links not only among the internal enterprises but also with other types of industrial/economic/institutional agents operating in the same geographical area/region. This is the approach of the studies on the Italian industrial districts due to Becattini (1990), who developed the original ideas of Marshall (1950) and gave a sociological definition of this type of cluster in the following terms: "a socio-territorial entity which is characterized by the active presence of both a community of people and a population of firm in one naturally and historically bounded area". The idea is that the role of agglomeration and proximity of firms is not the unique condition promoting an effective clustering: the attitude to cooperate of the local population is important as well, together with a good division of labour through its dissemination to internal enterprises.

The sociological analysis has directed the attention of researchers and managers towards new characteristics of a cluster to take into account: the presence of a *social capital* that could strongly condition the development of the enterprises and then of the cluster, and the presence of typical mechanisms of internal knowledge production and dissemination. In the Italian industrial system, the social capital is an important basis of any district. An industrial district, indeed, has often been considered a community of persons belonging to the same social environment, then sharing the same culture, history, ethical and human values. Perhaps, this could sometimes be considered a vision of the recent past, because today things are changing due to involvement of new people also coming from foreign countries; however, a common vision of the labour ethics and common behavioural rules still surely exist. By using a Marshall concept, in a SME balanced cluster as an industrial district is, benefits of a special and diffused "industrial atmosphere" can be recognized, both in terms of social capital, with an enriched set of interactions among internal agents, and of consideration and diffused thrust.

Sociological analysis reveals that the success of the Italian district model also comes from a local presence of industrial knowledge derived from the artisan origin of the internal SMEs, that is, the knowledge from which the local technological competence afterwards applied in the enterprises has been derived. Then, a SME cluster can be represented as a "knowledge system", that is, a set of agents where specific processes of knowledge production and dissemination, typical of the cluster region, exist. In several SME clusters an internal learning mechanism, in individual SMEs, occurs and is favoured by both the high labour specialization and division, and by the greater opportunity of personnel to change employment from one SME to another (as some districts analysed in the CODESNET website have shown).

The SME cluster characteristics emphasized by the sociological approach, identified by several researchers in analyses particularly developed in some Italian districts, surely appear to be significant conditions for a robust cluster organization (Antoldi 2006). However, it seems to be very difficult to identify them, and mainly to evaluate how they can be transferred in new clusters, or improved in existing ones, and in general how large their importance is and how long they could be maintained. All these questions need to find some answers, as general as possible, and their relative importance should find an as formal as possible evaluation: only in this case, could their real effects be estimated and their dissemination in new and old clusters assured. This problem, that seems difficult to approach, however needs to be approached, which is then a further motivation for the CODESNET-driven research.

The third analysis approach considered, namely the industrial approach, has modelled an industrial cluster as a complex production system, composed of several production units, often with high specialization, interacting together through a capillary network of logistic and information transfers, located in a specific region. On one side, several enterprises, of different nature and possibly of not so different dimensions, all belong to a common industrial sector; on the other side, there are a number of social, either private or public agents, such as schools, research institutes, service agencies. The scope of this further analysis approach is mainly to estimate the behaviour of the internal enterprises to understand how their inclusion in the cluster could increase their profit, as well as to evaluate the coordination of the cluster in order to know its effectiveness. With reference to the Italian case, this is the goal of studies developed by the research offices of the Bank of Italy (Banca d'Italia 2004) and others.

Indeed, the main goal of this last approach is to prove that SME clusters are enterprise agglomerations regulated by *interactive communication* and by *collaborative management*: the first characteristic assures the reciprocal knowledge that each SME management needs to have in order to adopt proper strategies not conflicting with other SMEs; the second characteristic refers to the agreements, reciprocal constraints and thrust situation which will guarantee that each individual management strategy will be generated such to favour the cooperation and to keep the competitive desire low. But a very complex management is today a real practical as well as theoretical problem: it needs more and more investigation, both on these cluster characteristics and on their dynamics. Then, the CODESNET approach has also been focused on this aspect.

Indeed, just the last two concepts, the interactive communication and the collaborative management among the internal SMEs, represent the two keystones of the methodology, which has been developed in the aforementioned EU-funded project, and here presented.

Interactive communication implies the utilization of a network of connections for transferring information, materials and persons among the enterprises internal to a cluster, as well as from these internal enterprises and the external markets. To analyse such a network, with evident dynamical characters, it thus requires the use of a model, as formal as possible, able to describe different kinds of interactions among autonomous agents.

On the other hand, *collaborative management* is a typical concept concerning industrial business, that refers to how existence of interactions among different enterprises usually push each individual management to search for agreements which would promote collaborative strategies for production, design and marketing among enterprises themselves, but still guaranteeing a local autonomy, even if partial.

The analysis approach presented in the following chapters, by translating the above two concepts into a formal model of an industrial cluster, will give an integration of the three traditional approaches found in the literature:

- The economic approach. This will inform the modeling of the network of interactions among the enterprises.
- The sociological approach. Its concepts will drive the modeling of the exchange of information and knowledge among the internal enterprises.
- The industrial approach. This last one will suggest how to model criteria and methods to manage collaborations among the enterprises.

This integration of the three approaches, and related cluster representation/modeling viewpoints, and its translation into a formal model of an industrial cluster will be the real novelty of the CODESNET approach presented here.

References

Albino V, Kühtz S (2004) Enterprise input–output model for local sustainable development. The case of a tiles manufacturer in Italy. Resources Conserv Recycl 41:165–176

Antoldi F (2006) Between local tradition and global competition: introduction to phenomenon of Italian industrial districts. In: Antoldi F (ed) Small Enterprises and Industrial Districts. Il Mulino, Bologna

Asheim T (1994) Industrial districts, inter-firm co-operation and endogenous technological development: the experience of developed countries. In: The United Nations (ed) Technological Dynamism in Industrial Districts: An Alternative Approach to Industrialization in Developing Countries. The United Nations, New York

Banca d'Italia (2004) New research of the Bank of Italy on Territorial Development (in Italian), Rome

Becattini G (1987) Market and Local Forces: the Industrial District (in Italian). Il Mulino, Bologna

Becattini G (1989) Sectors and/or districts: Some remarks on the conceptual foundations of industrial economics. In: Goodman E, Bamford J (eds) Small Firms and Industrial Districts in Italy. Routledge, London, pp. 123–135

Becattini G (1990) The Marshallian ID as a socio-economic notion. In: Pyke F et al. (eds) IDs and inter-firms co-operation in Italy. International Institute for Labour Studies, Geneva, pp. 37–51

Bergman EM, Feser JE (1999) Industrial and Regional Clusters: Concepts and Comparative Applications. In: Jackson RW (ed) Web Book of Regional Science. Regional Research Institute, West Virginia University, Morgantown, WV. http://www.rri.wvu.edu/regscweb.htm

Brenner T (2004) Local Industrial Cluster: Existence, Emergence and Evolution. Routledge, London

Brusco S (1982) The Emilian model: productive decentralisation and social integration. Cambridge J Econ 6:167–184

Bullinger HJ, Kuhner M, van Hoof A (2002) Analyzing supply chain performance using balanced measurement system. Int J Product Res 40(15):3533–3543

Caunce S (2007) Revealing A New Northern England: Crossing the Rubicon with Daniel Defoe. Prose Studies 29(1):136–152

Defoe D (1724) A Tour Thro' the Whole Island of Great Britain, 1971 edn. Penguin, London, p. 10

Everitt A (1979) Country, County and Town: Patterns of Regional Evolution in England. Trans R Historical Soc 5th Ser 29:79–108

European Cluster Observatory October (2007) Innovation clusters: a statistical analysis and overview of the current policy support, http://ec.europa.eu/enterprise/newsroom/cf. Accessed 2007

European Commission (2002) Regional Clusters in Europe, Observatory of European SMEs 3, Enterprise Publication, Brussels. http://europa.eu.int/comm/enterprise

European Commission (2003) SME and cooperation, Observatory of European SMEs 5, Enterprise Publication, Brussels. http://europa.eu.int/comm/enterprise

Fabiani S, Pellegrini G (1998) Un'analisi quantitativa delle imprese nei distretti industriali italiani: redditività, produttività e costo del lavoro. L'Industria, XIX (4):811–831

Gaggio D (2007) In Gold We Trust: Social Capital and Economic Change in the Italian Jewellery Towns. Princeton University Press, Princeton, NJ

Granovetter M (1985) Economic action and social structure: the problem of embeddedness. Am J Sociol 91:481–510

Haddadi A (1995) Communication and cooperation in agent-systems: A pragmatic theory. Springer, Berlin Heidelberg New York

INTEROP (2000–2008) http://www.interop-noe.org. Accessed 2007

Lazzeretti L (2002) Co-evolution and financial relationships between the banking system and the local industrial community of Prato industrial districts (1936–1999): An ecological approach. In: Working Paper (Small Business) 13, University of Birmingham, Birmingham

Lazzeretti L (2003) Density Dependent Dynamics in Arezzo Jewellery District (1947–2001): Focus on founding, presented in the Regional studies Association International conference Reinventing Regions in a Global Economy, Pisa, Italy 12–13 April 2003

Love J, Roper S (2001) Location and network effects on innovation success: evidence for UK, German and Irish manufacturing plants. Res Policy 30:643–661

Marshall A (1950) Principles of Economics, 8th edn. MacMillan, New York

Molina-Morales FX (2001) European industrial districts: Influence of geographic concentration on performance of the firm. J Int Manage 7:277–294

Nohria N, Eccles RG (eds) (1992) Networks and Organisations: Structure, Forms and Actions. Harvard Business School Press, Cambridge, MA

Paniccia I (2002) Industrial districts: evolution and competitiveness (in Italian) firms. Elgar, Cheltenham

Paniccia I (1999) The performance of IDs. Some insights from the Italian case. Human Systems Manage 18:141–159

Picard PM, Toulemonde E (2003) Regional asymmetries: economies of agglomeration versus unizoned labor markets. Regional Sci Urban Econ 33:223–249
Rosenberg N (1983) Inside the Black Box: Technology and Economics.
Cambridge University Press, Cambridge
Rosenfeld S (1995) Industrial strength strategies: regional business clusters and public policy. Aspen Institute, Washington DC
Silijak DD (2007) Large-scale dynamic systems: stability and structure. Dover, Mineola, NY
Villa A (2003) Some design criteria for a manufacturing virtual enterprise, Int J Auto Technol Manage 3:173–184
Villa A, Antonelli D, Cassarino I (2005) Issues in the management of collaborative demand and supply networks. In: AAVV. Strengthening competitiveness through production networks. European Communities, Belgium, pp. 47–57

Chapter 2
A View of SME Clusters and Networks in Europe

Abstract It is a matter of course that each country in the large European Union presents specific characters and individual features of its own industrial environment. However, a common peculiarity can be recognized, evidenced by two numbers: the percentage of SMEs in any national industrial system, always close to 90% of the total number of enterprises, and the percentage of personnel employed in SMEs, greater than 60% of the active population. What can also be widely recognized in almost all European countries are the recent crises, which have affected SMEs, and the attempt by SMEs to counteract their difficult position by searching for agreements and cooperation. One type of reciprocal support SMEs looked for in a crisis was contracts with larger enterprises: this gave rise to supply chains. But often the desire of SMEs was to have collaborative links with other SMEs, operating in the same industrial sector and mainly located in the same region: this resulted in the rise of networks and districts. In the last decade, the European Commission has started to promote studies devoted solely to supporting these types of clustering. Some countries have also launched programs to finance SME aggregations, defining agencies for pushing the establishment of new SMEs groups. This chapter offers an outline of a number of different national situations, concerning the rise and, sometimes, the fall of SME clusters and networks. Obviously, the scope of this chapter is not to give an exhaustive presentation of the European situation of SME aggregations: it aims to force the reader to recognize similarities, weakness and strength aspects, and to apply these to an analysis of the SME aggregations performance.

A. Villa, D. Antonelli, *A Road Map to the Development of European SME Networks*,

2.1 An Overview of SME Networks Across Europe

T. Potinecke and T. Rogowski – Universität Stuttgart

In the European industrial system a network of small to mid-sized enterprises (SMEs) usually consists of individual firms, separate business units, project teams or groups of organizations, which are formally or informally connected in order to exploit synergies. Cooperation agreements can be temporarily stipulated, either over the short or long term: at the end of the cooperation agreement in the network, each SME will operate again independently.

Considering an individual SME, which is considering either to participate in a network or operate in a market of continuously growing competitiveness, the decisive difference, if the former option is adopted, refers to the condition of subordinating main individual plans to the network collective scope. This main characteristic of the networked organization justifies why networking advantages for SMEs come from reduction of transactions costs among the network components, as well as a more stable coverage of markets owing to reciprocal trust (Thomson 2003).

Networks cannot be reduced to either formal or informal connections of different units. The qualitative aspect of networks as "learning or knowledge communities" is one of the most effective competitive advantages in the long-run. This leads to the question if knowledge networks are more effective than institutions. Indeed networks are a kind of institution with habits, norms, rules and routines, but they have the ability to adapt themselves to changes faster than institutional organizations. Knowledge networks are mainly influenced by research and innovation activities which need flexible structures to overcome environmental dynamics. Then, networks of innovative firms, which co-operate in R&D processes – like product and process development activities – in a delimited area appear to be of particular importance in the European context.

From existing experiences, the organization of an industrial network through progressive grouping of some SMEs can occur according to some steps.

The first step of each network constitution/development process is the idea and impulse initialized by promoters, because of the need or lack of resources for reaching the foreseen benefits. The second step is the construction of the cooperation agreement and the networking rules (constitution phase). An important factor during this crucial step is the selection of the partners based on an intensive relationship management. The third step can be seen as the networking stabilization ("work in the network"), characterized by the assessment of the form of work, products, services, costs and risks (e.g. saving finance, reaching aims and identification, "actualizing" trust and reputation or developing products), such to have criteria and internal regulations in place for crises and conflicts. Further issues are the development of short- to long-term relationships, contracts, rights and open structures. Step four is defined by the evaluation of the set targets' realization and is supported by monitoring processes (Podolny 1998; Williams 2000).

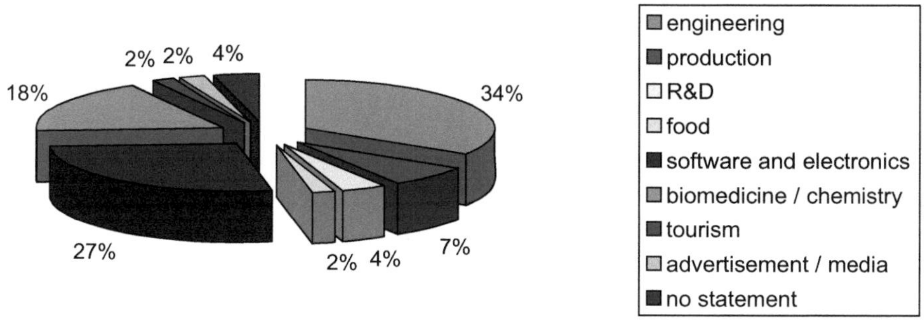

Fig. 2.1 European results concerning specialization

To have an overview of the experience of SME network organization in some European countries, the current situation of a number of clusters, concerning their success in terms of collaboration and main network attributes, was observed during the development of the CODESNET project. As seen from the CODESNET website (http://www.codesnet.polito.it) public data on more than 100 clusters from 11 different countries all over Europe have been collected and analysed. The sample includes mainly the engineering, software and electronic sectors, as well as some further clusters to complete the picture as shown in Fig. 2.1.

The data analysis, developed according to the CODESNET model of a SME network (as described in the Chap. 4), shows both unexpected as well as already detected results.

- Fifty-five percent of the clusters show a clear division of labour among the partners, thus making the coordination easier and more profitable.
- Only 35% of the assessed clusters can afford a dedicated supporting information and communication technology.
- There is no clear trend concerning the respective sales market, as well as no trend concerning the existence of a cluster-wide organization structure.
- Forty-seven percent of all assessed clusters apply conjoint marketing strategies and activities.
- Concerning the quantitative assessment, 46% show a high improvement potential and 51% can be found in the middle success category, whereas there is still potential to improve the collaboration or organization within the clusters.

There are different situations for geographical distributions of networks: on the one hand a network can be agglomerated in a relatively narrow area (rural district, county); on the other hand a network can consist of firms with a distribution over a whole nation or (but rarely) over Europe.

From public data, it is quite difficult to evaluate the financial situation of the clusters due both to the partial lack of relevant data and information, and to the effective difficulty in estimating the network economic data as opposed to that of the individual SMEs. This aspect, today, is an open problem and it has promoted specific analyses and studies, currently under development.

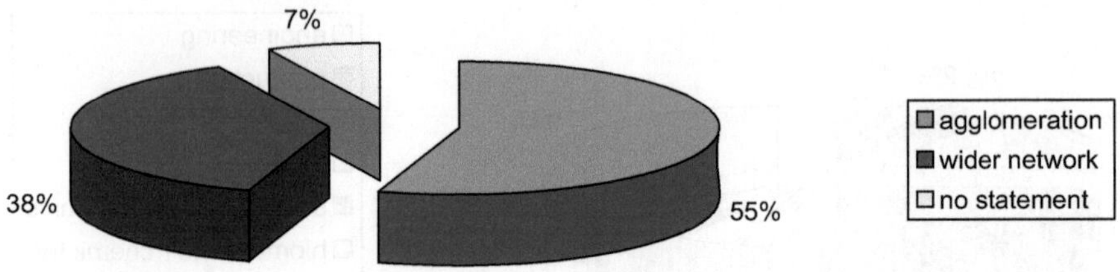

Fig. 2.2 European results concerning geographical distribution

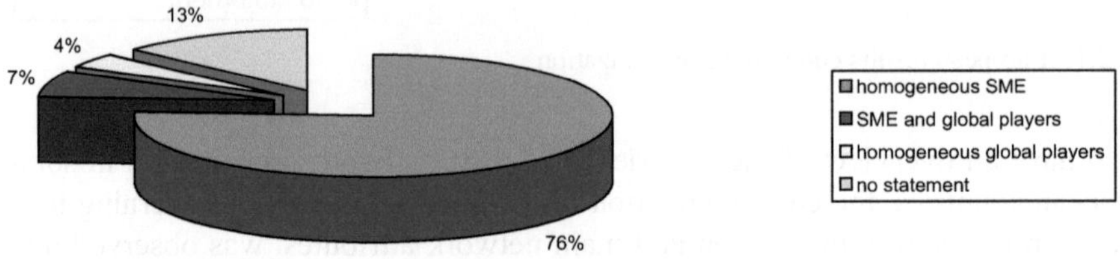

Fig. 2.3 European results concerning average firm dimensions

Referring to the set of collected SME networks during the CODESNET project, Fig. 2.2 shows that 55% of the evaluated clusters did not give any information about their sales rate. Accounting for the detected ones, the majority of SME clusters in Europe have sales between one million and one billion Euros.

To remark on the dominance of SME in European clusters, Fig. 2.3 offers an overview of the average firm dimensions. The majority of firms in the examined European clusters are SMEs (76%). The structure of the assessed clusters is mostly homogeneous concerning the firm size.

An overview of the most evident characteristics of SME networks and clusters in some European countries (where one or more institutions have been involved in the CODESNET project) is summarized in the following. A detailed description of the typical networks and clusters indeed requires a wider and wider presentation. However, the aim of this section is simply to outline the main features of the most diffused SME agglomerations: specific analyses will be reported on in the next parts of Chap. 2.

A study of mechanical engineering clusters in Germany shows that at least 227 agglomerated industrial areas exist with at least one mechanical engineering cluster. The 227 areas have about 542 specific clusters of special branches. Out of the identified 981 potential mechanical engineering clusters, only 55% can be considered a real cluster when applying the holistic definition. The highest concentration of mechanical engineering clusters can be found in the south of Germany as well as in the provinces of Thüringen and of Sachsen. Small specific agglomerations also exist in the south of the provinces of Nordrhein-Westfalen (area of Hamburg), as well as in the hinterland of Berlin. The results of the evaluation of German clusters collected by the CODESNET project show a medium sector diversification and identify the engineering sector as the most important in Germany. Concerning the distribution over a special region, no special trend was observed, since the networks and clusters are agglomerated and distributed nation- and European-wide to the same extent. The sales rate of the assessed German clusters ranges from 0.2 million to 36.7 billion €; the firm dimension in terms of employees from less than 20 up to over 10,000. In German clusters, a division of labour among the partners exists more often than not. Special Information and Communication Technology (ICT) is only applied by a minority, as well as only a minority established a cluster-wide organization structure. More German clusters are export-oriented rather than oriented towards the national German market (this situation goes hand in hand with the export orientation of the German mechanical engineering branch to which many clusters belong).

Concerning the SME networks in Italy, several "industrial districts" participated in the CODESNET evaluation. Compared to other nations, the clusters in Italy are diversified on the national level, but almost every one is specialized in a certain branch or niche. The allocation of nearly all clusters shows agglomeration and almost no European or national distribution. Sales reach from 1.1 million up to 6.5 billion €. Mainly SME with 15–200 employees are part of Italian clusters: only in one case the exception global player with over 10,000 was involved. In contrast to the German clusters, nearly all Italian clusters practice division of labour. As well as in Germany, almost all clusters in Italy do not use special ICT. The market orientation clearly indicates the national market as the main target. Trends, in either way, cannot be identified when concerning the organizational structure.

The clusters in Hungary, as assessed in the CODESNET analysis, are rather specialized in software and engineering services. The distribution varies from agglomerations in one county up to Europe-wide linkages. Although the clusters are rather specialized, a relatively high number of different fields of skill and knowledge are employed. Considering the firm size in terms of employees, the Hungarian clusters consist of rather small firms ranging from 5 to 50 persons. All clusters operate a distribution of labour. Due to the specification of software and immaterial services, no transportation means are needed and were therefore not indicated. A specific ICT does exist in any of the assessed clusters. Concerning the market orientation, the national market is more important than the export. Special organizational structures mostly do not exist, whereas the majority of Hungarian clusters are leading firms and have common marketing activities.

The clusters in Greece participating to the CODESNET evaluation are rather diversified on the national level, software and chemistry being the main sectors. The distribution of the respective networks is rather nation- and Europe-wide. Compared to other clusters in Europe, the partners of the Greek firms are rather small in terms of employees (ca. 8). Some of the analysed Greek clusters possess an organization structure with interesting characteristics, some other comprise leading firms and show conjoint marketing strategies.

Most of the Polish clusters are very specialized, on environmental technology and on advertisement and media. The distribution is rather in local agglomerations than nation- or Europe-wide. Two to three skill or knowledge fields are covered by the employees on average and the firm size varies from 1 to 2,000 employees. Division of labour as well as an applied organizational structure and conjoint marketing strategies exist in the most clusters. The market orientation leads rather into the direction of the national Polish market.

The French clusters vary in their diversification degree. On average several sectors are covered, whereas chemistry and engineering hold the main parts. Concerning the geographical distribution, the clusters are mostly agglomerated. Relatively homogeneous is the sales rate, from 250 to 600 million €, in contrast to the firm size which rages from 10 to 10,000 employees. Division of labour as well as special ICT for the respective cluster exist for the majority. In contrast to the existence of conjoint marketing strategies, which are mostly applied, a trend for the existence of an organizational structure is not recognizable. Furthermore the disposal orientation leads rather into the direction of the national market.

Concerning the United Kingdom, the CODESNET analysis has been focused on specialized clusters on aerospace and information technology. These clusters indicate nation- and Europe-wide linkages in their networks, instead of local agglomerations. The sales range from 60 million € up to 5.2 billion € per year while the number of employees in the cluster firms ranges from 20 up to tens of thousands. A clear division of labour and wide application of ICT have been realized. However, British clusters usually show a lack of organizational structures. Concerning the disposal market and the existence of larger leading firms in the clusters, no trend can be recognized by the available information. In contrast, a clear lack of conjoint marketing strategies and activities can be identified.

The clusters from Ireland analysed by CODESNET run across industry lines rather than being specialized. The geographical distribution of the respective firms is nation- and Europe-wide, mostly not agglomerated. Most important is the engineering sector (mechanical and electronic). Annual sales range from 6 to 400 million €, while the average firm dimension is around 10–25 employees. Almost all transportation means (trucks, ships and planes) are used by manufacturing clusters. Leading firms as well as conjoint marketing activities are lacking in the majority of the Irish clusters. In contrast to that, most of the clusters afford special ICT applications. A trend concerning the existence of an organizational structure is not recognizable and the disposal market is rather the national market than external ones.

Referring to the Swedish clusters available in the CODESNET catalogue, they are rather specialized, whereas they are diversified on the national level. Tourism and mechanical engineering build the balance point. All clusters are agglomerations and consist of homogeneous SME in terms of number of employees. Leading firms and conjoint marketing both exist in the majority of the clusters, whereas there is only one which owns an organizational structure. The national market is the favourite disposal market for the majority.

The Finnish clusters collected by CODESNET are relatively specialized, especially in software and Research and Development. They indicate an agglomerated distribution instead of wider network linkages. Despite the specialization, the clusters do not consist homogeneously of SME and none of them apply conjoint marketing activities or strategies. A trend concerning the disposal market is not recognizable as well as a trend concerning the existence of special ICT.

As mentioned before, the short list of comments above is merely a summary of considerations coming from an overview of the networks' catalogue collected by the CODESNET project. In principle, they can offer a concise description of the main issues, which can be seen in European countries where clusters and networks have a relevant presence in the respective industrial systems.

The following sections will address the main features and also weakness and strength aspects facing the SME networks in some countries. These aspects will highlight the most significant problems concerning network organization, their management and development, in the current dynamics of the European as well as worldwide markets.

2.2 Poles of Competitiveness in France

X. Boucher and A. Dolgui – École de Saint-Étienne

In the context of a global economy characterized by increasingly severe competition, France launched in 2004 a new industrial policy to develop key competitive factors, among which of primary importance is certainly innovation.[1] This new policy induces and supports initiatives emerging from academic and economic actors throughout a region to develop dynamic networks which link firms, and research and educational institutions. This program is based on the creation of an official and financially supported designation called "poles de compétitivité". To date 71 competitiveness clusters have been created, including 17 which already have, or will have, a worldwide impact.

The aim of a competitiveness cluster is to concentrate at one location the talent incorporated within public and private research laboratories, teaching facilities and business enterprise expertise, in order to establish working relationships which should develop a cooperative environment and promote partnerships within innovative projects. Thus, the definition of a competitiveness cluster is based on the following key points:

- An association of companies, private and public research centres, and teaching institutions
- Collectively involved in a public/private partnership (requiring a common development strategy)
- Aimed at launching new projects resulting in innovative technological and organizational advances, increased economic efficiency and job creation
- In principle, enabling those players involved to become leaders in their field

[1] All information presented here on competitiveness clusters is based on public information, notably made available by the French Ministry for Economy Finance and Industry.

Four key elements for successful clusters have been identified:

- A common *development policy* consistent with the overall economic strategy of the territory
- *Close partnerships* among the various players linked to well-defined projects
- *Concentration* of highly marketable *technologies*
- *International visibility* based of a sufficient critical mass

The French government provided 500 million € annually from 2006 to 2008, in financial support, mainly to the industrial R&D sector, which is at the heart of the competitiveness clusters policy. This monetary aid applies to all the partners involved, regardless of size or origin, and therefore will enable external international players to become involved in these projects.

2.2.1 The French Context and Advantages Offered by Competitiveness Clusters

By the creation and support of competitiveness clusters, France launches a high-impact industrial policy, expected to augment innovation capabilities and establish competitive advantages. By regionally networking all the actors in the innovation process, the cluster policy is expected to bring new sources of value creation and develop employment within the regions. Indeed, even if France possesses lots of existing skills and talents – notably a high level of creativity – the coordination of these strengths clearly appears insufficient, especially for economic actors with similar activity domains distributed over a limited territory.

This economic development policy confirms the industrial sector as a key driver of the overall French economy: industry represents the main source of innovation (90% of R&D expenses) and competitive factor (80% of exports). The industrial sector plays the role of catalyst for the rest of the economy. However French industry is confronted with a twofold challenge linked to the evolution of the global economy:

- *Globalization of exchanges and production processes* which increases the levels of competitive pressures
- *Establishment of a knowledge-based economy*, where innovation, research, i.e. immaterial capital or collective intelligence, turns out to be one of the main vectors of economical development and competitiveness

To answer such challenges, the competitiveness clusters intend to build new coordination facilities based on the operational interconnectivity among territorial development, innovation and the industrial sector. Hence, the networking of industrial, scientific, and education actors over a well-defined territory will constitute:

- Faster and broader innovation (proximity among actors stimulates information sharing and competence flows; helps the emergence of highly innovative projects).
- Larger exposure: the concentration of participants in a limited territory offers a national and international visibility thus attracting more and better players.
- A counterbalance to economic outsourcing (competitiveness is directly linked to the implantation of enterprises within the territories, induced by the presence of pertinent competences and useful partners).

By highlighting innovation and the sharing of common savoir faire, such clusters will improve internationally the competitive attractiveness of the French territory, and will promote the regional specialization of French industry, as well as assist the emergence of new activities with significant international visibility. As a synthesis, the purposes of competitiveness clusters are:

- To align French creativity towards value creation
- To develop the French economy and competitiveness over the long term
- To concentrate highly technological activities in selected territories
- To heighten the economic attractiveness of French regions
- To improve employment and to limit outsourcing

2.2.2 Thematic and Geographical Distribution

The latest decisions of the French government have created a total of 71 competitiveness clusters (Fig. 2.4). These clusters concern emerging technological sectors such as nanotechnologies, biotechnologies or microelectronics, but also are involved in the more traditional industrial sectors. Among these 71 clusters, seven have the ambition/aspiration to become world-class leaders. Ten other clusters are slightly smaller, but aim to become also world-class leaders over time. The 51 remaining clusters have national and regional objectives, and among them 36 are applied to the industrial sector.

Different types of partners can apply to the financial support associated with the clusters, regardless of size or origin. This is especially important since it will allow international players to take part and profit from these projects. Welcoming outsiders is a significant part of diversifying the industrial and economic landscape within a region or in France overall. Moreover, this synergy also applies to the public sector, since the goal of these competitiveness clusters is to concentrate the efforts of the French government, national agencies and local authorities.

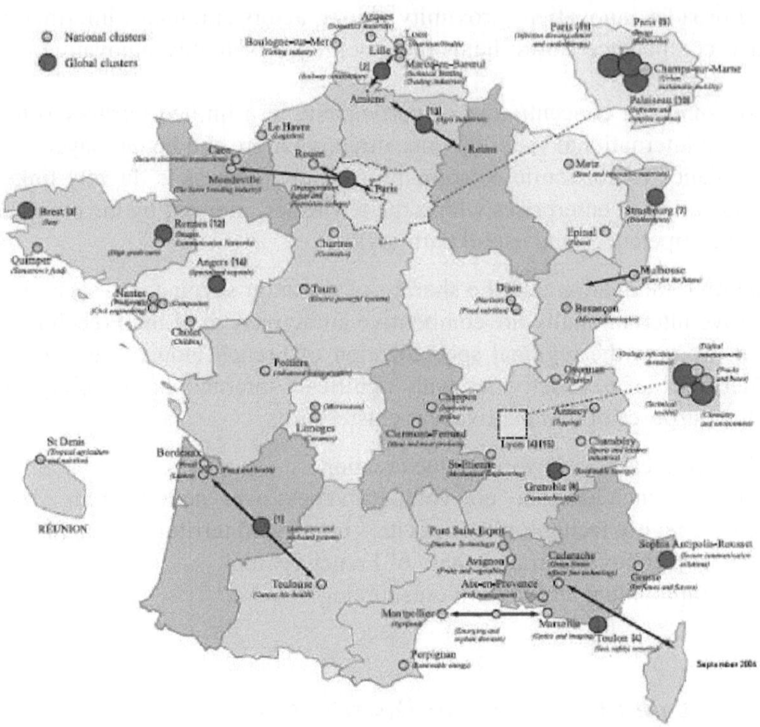

Fig. 2.4 Geographical distribution of clusters edited by the French Ministry for Economy Finance and Industry

2.2.3 *Governance and Managerial Mechanisms*

2.2.3.1 Governance Structure

The competitiveness clusters are managed in a project-oriented scheme. Clusters should become innovative sources of collective projects among enterprises, research centres, and training institutions.

- Research and development (R&D) projects are the core activity for these clusters, and the main competitive factor.
- Non-R&D projects (training, infrastructure investments, IT investments, economic intelligence, international development, territorial development, etc.) constitute an indispensable supplement to ensure sustainable economic development and competitiveness.

Each cluster is to be represented and animated by a *coordination entity* with a specific legal structure, usually based on the status of an association. This governing body has to ensure a prominent management role to the industrial, academic and scientific actors, and a lower but clear representation for regional institutions. The coordination entity employs a permanent staff whose mission consists mainly in playing the role of the key facilitator in the emergence of innovation projects among the various actors concerned. National and regional governments both contribute to the funding of the clusters.

The principal mission of such a coordination entity consists in:

- Elaboration and deployment of the general strategy for the cluster
- Coordination and selection (with a designation) of research projects which can become candidates for public funding assigned to clusters
- National and international communication in the cluster
- Launching cooperation with other clusters in France and abroad
- Evaluation of projects

A mid-term program is defined to ensure the adequate supervision of the relations among the clusters, government, and territorial institutions.

2.2.3.2 Financial Mechanisms

An important financial incentive has been offered to guarantee the success of the competitiveness clusters. The national government has planned a budget of 1.5 billion € over three years, with three main forms of financial aid:

- First, a project-oriented budget is constituted with contributions from several complementary ministries (Industry, Equipment, Defence, Agriculture, Territorial Development and Competitiveness (DIACT)) and from regional development agencies (Agence Nationale de la Recherche, Agence de l'Innovation Industrielle, Agence OSEO for Innovation in SME). A major portion of this budget is directly destined for R&D and innovation research programs, which are at the heart of the clusters.
- Second, the company and the employees implicated in R&D programs validated by the clusters can take advantage of tax exemptions and reductions of social charges.
- In 2006, the French government created 3,000 positions in the research sector: the major part of these posts concern the scientific domain covered by the clusters.

This national budget gathered from several distinct sources has been centralized within a unified fund so as to simplify its administration. Supplemental funding will come from other territorial institutions, like the regional governments which also have a clear mission to provide support to competitiveness clusters. These institutions already supply substantial amounts in addition to the national budget.

2.2.3.3 Direct Help for R&D Projects

The use of this budget is submitted to a systematic selection process: R&D collaborative projects from all areas covered by the competitiveness clusters are analysed, then selected on the basis of equitable and transparent decision criteria. Calls for projects occur three times each year. The major selection criteria are:

- Concrete results in terms of value creation, economic activity, and employment
- Innovative technological content
- Development of new products and services with a market focus
- Consistency among the projects goals, the global strategy of these clusters, and the local strategies for companies

A custom governance structure has been defined to ensure the selection and supervision of the projects. A typology of projects was defined, with a specific agency in charge to supervise each of the four types of projects, as defined in Table 2.1.

Table 2.1 Typology of projects and the supervising agency

Project type	Supervisory institution	Selection procedure
Fundamental R&D project, where there is a technological gap, with long-term view on market deployment	Agence Nationale pour la Recherche (ANR)	– Periodic call for projects – Priority given to cluster projects
Industrial collaborative R&D projects with mid-term market impact (five years)	Ministry for Economy Finance and Industry	A ministry expert nominated for each cluster
Individual SME innovation projects	OSEO-ANVAR Agency	– Permanent call for projects – Priority given to cluster projects
Very ambitious and long-term innovation projects (budget > 10 millions euros)	Industrial Innovation Agency	Specific expertise to support the definition and management of such projects

2.2.3.4 Tax Exemption Opportunities

Tax exemption can only be applied by a company implanted in so-called R&D geographical areas linked to the clusters, and participating in one or several R&D projects. These exemptions concern both national and regional taxes. For the employees directly participating in the projects, social charge exemptions can be granted. The amount depends on the size of the company: 50% of exemption for SMEs, and 25% for remaining companies.

2.2.4 Case Study

ViaMéca is the French Rhônes-Alpes cluster in the field of the mechanical industry (http://www.viameca.fr). Industrial groups, SMEs, research centres and partners, all have plotted a common future with the objective of making the ViaMéca cluster a leader in the mechanical engineering sector with global visibility. ViaMéca intends to play the role of an accelerator by creating new relationships between the players in the mechanical engineering sector. By putting the development of the mechanical engineering industry at the heart of its strategy, ViaMéca is weaving a network of businesses with multiple, complementary skills. With only three years of existence, already 20% of the French mechanical engineering companies have come together under this designation.

2.2.4.1 Industrial Context and Purpose

In an international context of competition globalization, the manufacturing and mechanical industry has to shift towards a highly innovative mode of operating, while maintaining the current objectives of lean production and cost reduction. For such challenges, marketing and technological innovation require to be associated with an increased level of reactivity among all stake holders. Technological as well as organizational innovations will constitute a necessary support to implement profound transformations of production and design modes. The Digital and Intelligent Manufacturing Plant will become a reality, by integrating tools from both the world of machines and the realm of data acquisition/treatment. Mechatronic production lines provide a successful and excellent example of such integration, with intelligent sensors embedded at the heart of the manufacturing devices. The power of innovative manufacturing processes integrated with the ubiquity of information management will constitute one of the major competitive factors in the near future, supporting both product and organization innovations.

However such progress will only be possible if it is associated with complementary actions: significant increase in vocational training, consideration of new environmental and sustainable development factors, notably induced by new constraining French and international regulations.

ViaMéca competitiveness cluster puts the focus on the deployment of global integrated innovation and engineering (including product design, manufacturing innovation, life cycle management, etc.). This overall integrated innovation and engineering is considered as a weapon against the risk of regional industrial activities being outsourced. This new paradigm makes it possible to compete on the global market with new arguments, distinct from the old quality-cost-delay triangle, and based on novel differentiations of products and associated processes. By putting the focus on research and innovation, this general and integrated vision of the mechanical industry will also require conserving viable local manufacturing potential, to make possible the emergence and testing of prototypes and innovations.

Table 2.2 Key figures

9,000 Businesses
235,000 Employees (20% of the French workforce working in mechanical engineering)
2,500 Researchers
Over 1,000 engineers or research Masters awarded diplomas each year

The Rhone-Alpes region has key advantages for such an endeavour. With a high density of industrial companies, as well as research and education institutions, the region has already launched such innovation-oriented dynamics in recent years (see Table 2.2). The *raison d'être* of the cluster is to finish the complete alteration from a region characterized by old industrial techniques, to become an innovation catalyst, forging the new manufacturing industry of the twenty-first century.

2.2.4.2 The Actors of the Competitiveness Cluster

ViaMéca cluster is supported by an essential mass of industrial companies as well as research and educational institutions (Fig. 2.5). The development of regional competencies remains a major goal, based on an intensive strategy of networking. To put forth actor networks, technological platforms will be developed, aiming at coordinating actors and enhancing industrial transfer from research (the regional

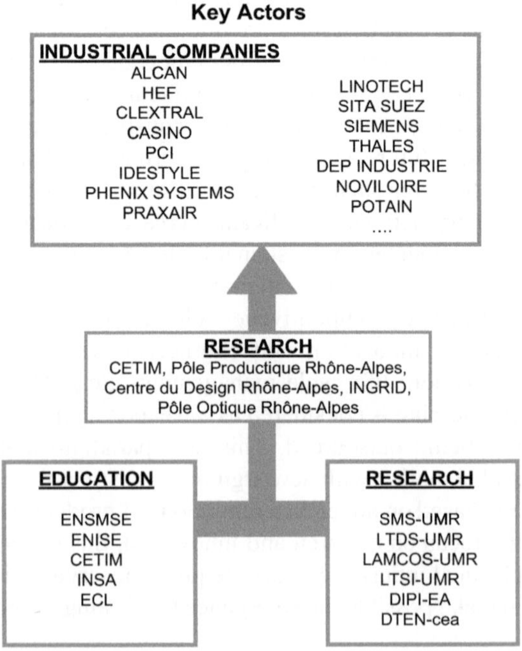

Fig. 2.5 Key participants of the ViaMéca competitiveness cluster

INGRID platform is a good example). The high density of the actors and the wide panel of technologies and processes in the region provide an excellent fertilizer for technological transfer.

Priorities have been given to three major technological domains: material sciences, advanced manufacturing processes, and design of products and sub-products.

These three fields are to be embedded within a global vision of integrated innovation, design and engineering.

In this region, more than 32% of industry works in the field of mechanics. There are 6,800 companies, with 550,000 employees (in 2003), which represents more than 20% of the whole industrial workforce. The mechanical area represents 16.8% of the added value of the regional industry with a turnover of 84.4 billion €. In the field of industrial equipment production, the Rhônes-Alpes region is ranked first nationally. This activity gathers 17.6% of the national employment in the same sector (57,640 employees) and produces 14% of the overall national product. The region has a second area of excellence with the metal finishing industry, which concentrates 16.7% of the overall national employment in this sector (55,180 employees).

2.2.4.3 Coordination Mechanisms and Strategy

The strategic committee initially created to launch the project has been transformed into an administration council. It is in charge of defining the overall strategy for the cluster and supervising its activities. The administration council works with an operational office in charge of day-to-day administration. At a second level, the cluster has defined seven complementary commissions to oversee specific coordination missions over the broad region (R&D activities, educational issues, etc.). These commissions are notably in charge to launch and coordinate the calls for projects linked to the cluster strategy. The third level of governance is at the project level: each project, linking various industrial, education and research actors defines its own coordination mechanisms. The strategy of the ViaMéca cluster is mainly oriented on the following topics:

- Development of innovation by the completion of R&D projects associating research centres, technical centres and industrial businesses
- Integration of (organizational or technical) innovation in industrial businesses and development of skills via collective actions
- International exposure, by setting up technological partnerships abroad

This strategy must be implemented over a long-term horizon, through three main phases. The first strategic phase follows a key objective of acceleration of technological transfer: making the regional businesses aware of cliental expectations; strengthening SME response capabilities as regards to innovation, quality, cost and delivery time; creating clusters or groups of businesses by speciality, thus ensuring complementarities of means of response (studies, development work, installations); initiating and participating in R&D projects by connecting public and private research laboratories.

The second strategic step is oriented on improving the global visibility of the region. Thanks to ViaMéca, SMEs/SMIs have access to major European contractors. ViaMéca will embark on an international development strategy by approaching European poles and clusters, by being complementary in their areas of activity, by facilitating the access of businesses to European projects, and by participating in the Mécafuture Platform.

The last phase will finalize this international development by obtaining a world-class industrial and scientific reputation. In that perspective ViaMéca emphasizes three key issues: obtain an international leadership for breaking technological deadlocks concerning specific domains of excellence; increase by 20% the number of researchers working on the cluster orientations and scientific publications; augment significantly the number of industrial innovations (innovation projects, funding and patents).

2.3 Science Parks in Greece

S. Agoti, C. Stylios and P. Groumpos – Patras

Science and technology are vital not only for the progress and the exploitation of knowledge, but also for the achievement of viable and balanced growth, stability and prosperity. Contribution of technology to economic growth and competitiveness is significant and is of great importance to innovation in any economy (Mowery and Rosenberg 1989). Science and technology parks (STEPAs) are essential means for transferring the scientific and technological knowledge from the research institutes and universities into enterprises. A STEPA is a mediator that contributes considerably to the regional growth, facilitating the creation of high-technology spin-off companies and disseminating innovative technological achievements to regional SMEs. For these reasons STEPAs all over the world are founded near universities and research Centres and are closely connected with them.

A science park is an organization focusing on the concentration of high-tech, science, or research-related businesses. Science park establishments host research-oriented SMEs and/or R&D sections of bigger enterprises. Science parks attract such firms because they provide various facilities. Their appeal comes from the neighbouring research and academic organizations and the offered infrastructures. Science and/or technology parks accommodate enterprises that produce commercial applications of high technology, exploit recent research results and use novel approaches in the sale and technical support of products and services (Chan and Lau 2005). Science and technology is based on the combination of research and innovation with the relative production and commerce.

Science parks' geographic proximity with research institutes, could be viewed as "the generation of new and valuable knowledge through human intervention" to the extent that an "innovative milieu", which generates constant innovation, is created and sustained (Hall 1994). It is proven that a science park incubator is recognized as an effective support mechanism for new entrepreneurial firms and its contribution is based on the framework of shared facilities such as offices, administrative staff and access to university research and external grant support from the government and other sources, such as venture capital (IASP 1998).

2.3.1 STEPA Characteristics

A STEPA is actually a complex economic and technological unit that aims at encouraging the growth and the implementation of high-technology and innovation production (Dettwiller et al. 2006). STEPAs provide services such as high-tech research installations, pilot laboratories, centre of innovation, centre of technology transfer, "incubators" for new firms, innovative techniques unit, etc. Most STEPAs today focus their activity on information technology (electronics and computers), telecommunications, biotechnology and new materials (IASP 1996).

The general characteristics of a STEPA are the following:

- Promotes and facilitates the transfer of research results from universities and research centres to the industry and more generally to the productive sector, increasing the economic growth
- Facilitates the creation and the viable growth of new innovative enterprises (incubator of enterprises)
- Constitutes a body of research and development for SMEs and new markets
- Provides the suitable environment where the knowledge-based enterprises can develop stable collaborations with concrete research and technology centres for reciprocal benefit

In economically developed countries, a STEPA establishment creates the environment and the conditions so that the whole region grows in new and different directions. In developing economies, the expectations and the role of STEPAs are "catalysts for growth". STEPAs facilitate the establishment of new high-technology companies and encourage the production of innovative products and services. Beyond the economic status of a country growth, STEPAs contribute effectively to the concretization of functional objectives that conform with the strategies of regional policy.

The total number of incubators for enterprises and science parks in Europe is roughly 900, which create roughly 40,000 new work places per year. The period of incubation (average period hosting an innovating company) is usually two to three years. Greece today allocates one incubator of enterprises per 106,000 SMEs. The corresponding number in Austria is one incubator per 3,000 companies (see Fig. 2.6). The minimum initial capital for starting a company in Greece is

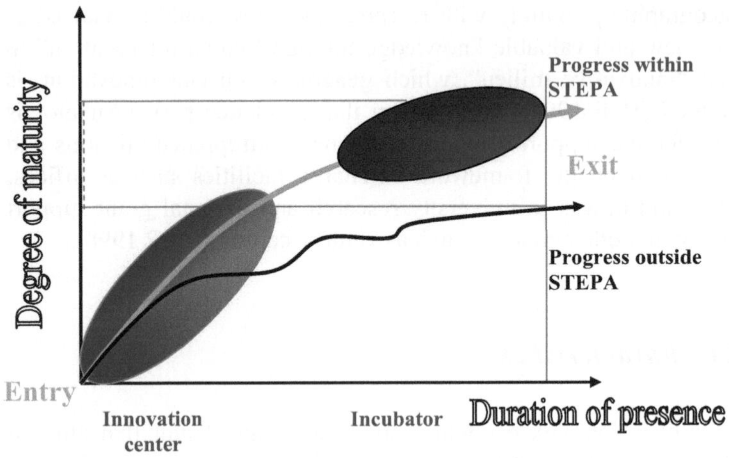

Fig. 2.6 The growth of a start-up company hosted in the innovation centre and then in the incubator of a science park

among the lowest in the European Union (0.017% of the GNP compared to the 0.036% of the European Community average).

The universities and the research institutes continuously propose and develop new technologies adopting innovative approaches. Even though a huge experience is accumulated and acquired, technological and innovative value remains as property, usually unexploited, by the researchers that worked in the particular programs. Moreover, there is not particular preparation of how the knowledge and all the innovative results will be disseminated to the enterprises. The difficulty in diffusing the research and technological results is more difficult than their usage for production.

In this direction, the role of STEPAs is essential. Their mission is mainly to bridge the gap between academic society (universities, research centres, etc.) and industry. In other words, the main role of STEPAs is their activation as "lighthouses of knowledge" for the diffusion of innovation and technology, so as the industry can directly use part of the enormous available scientific knowledge. Thus, they contribute effectively to the fast transformation of innovative results of research and technological development in successful enterprising undertakings.

2.3.2 Science Park Institution

In Europe, the institution of science parks is essential for the regional development and the general economic growth. One of the biggest science parks in Europe, and biggest in England, is Cambridge Science Park, which was founded by the Trinity College in 1970. It attracts relatively small local R&D enterprises directed to the new technologies. The presence of University of Cambridge, which has a great tradition in high-technology industry, supported the activities and the blooming of Cambridge Science Park. Many high-technology companies were founded in computer, information technology, telecommunications, software development, biotechnology, robotics, and the like. In 1970 it hosted 20 high-tech companies with roughly 1,500 workers. Until 1980, 360 companies were set up and 10,000 persons worked. Today, more than 1,700 high-tech enterprises with 40,000 workers have been developed globally (http://www.cambridgesciencepark.co.uk). Another best case science park in Europe is Lindholmen Science Park, in Göteborg. It was founded in 1995 in old shipyards, which had closed and 4,000 places of work were lost. Today, 10,000 people work in the Lindholmen Science Park. It hosts very important and well-known companies. The main activation areas are: mobile data communications, intelligent vehicles and transport systems and media (http://www.lindholmen.se/ext/index_en.php). In France, the most successful science park is Sofia Antipolis Science Park, which was founded in 1969. It includes 150 intermediate and big enterprises around the world. Main hosted companies are Dow Chemical, Digital Equipment, IBM, Cordis (US companies), as well as other Japanese, British, and Swiss companies. The Sofia Antipolis is the only park in Europe, which is not close to any university or technical university and in its early phase was supported by informal networks of partners that contracted initially a not-speculative company (http://www.sophia-antipolis.net). In Ireland, the Shannon Development was founded in 1959 and is situated in the area of University of Limerick supporting a network of science parks in the whole country (http://www.shannon-dev.ie).

2.3.3 Present Situation

The level of starting companies in Greece is among the lowest in the European Union. On the other hand, many universities, research centres and knowledge creation organisms are active in Greece, proposing new methods and approaches for products, services or processes but without being able to introduce their products to the market. Greece is mainly a rural area, where knowledge and technology productivity and new innovative creations need to be imported in the Greek market. For these reasons and in order to establish a communication channel between enterprises, universities and research institutes, the Greek government encouraged the inauguration of science parks.

The placement and distribution of science parks in Greece was done under the basis of the regional development advancement. Each Greek region presents different rates of growth, productivity, extroversion, unemployment and general social and economic status. In every region a science park was settled. The preparatory actions for the foundation of science parks in Greece dates back to 1988 and the first science park operated in 1989 in western Greece.

Western Greece was an industrial area, with local enterprises presenting huge rates of growth. In addition, the port of Patras, the capital of western Greece, was the gateway to European countries and to the West. In parallel, the University of Patras, the Technological Institute of Patras and other research institutions create knowledge and have many experienced researchers. Western Greece was blooming, until the 1980s, when all the huge industries closed, increasing unemployment and presenting negative growth rates. Consequently, the government decided to create a link between research bodies and enterprises, so that entrepreneurship and growth would be supported.

Patras Science Park S.A. (PSP) was established in 1989 as Patras Technological Park S.A. In 1992 it took its current name, while in 1998 it completed its premises at Platani, Patras. Its only shareholder, since 2001, is now the Greek State (under the supervision of the Ministry of Development and particularly GSRT).

Patras Science Park is an organization of a particular structure, establishing mechanisms and services primarily targeted to promoting the creation, operation and growth of "innovative firms". PSP contributes to the creation, operation and development of spin-offs based on innovation, and promotes their activities. Patras Science Park supports the completion of innovative ideas, products, services and procedures as well as the exploitation of research and development results, it encourages the constructive collaboration of knowledge creation organizations (KCO) and research institutes with enterprises, it promotes the introduction of new organizational and administrative methods for enterprises, the acquisition and diffusion of new knowledge for enterprises or any kind of legal entity and the attraction and installation of entrepreneurial schemes in the Park's premises (http://www.psp.org.gr).

The second try for the establishment of such an organization took place in Athens, the capital of Greece (1992). In the wider region of Athens, two science parks have been established, each one serving different purposes. The first is Lefkippos Technology Park (LTP), on the premises of the National Centre of Scientific Research "DEMOKRITOS" (NCSR "D"). The initial aim of the NCSR "D" was the utilization of the advantages of nuclear energy for peaceful aims. The centre has eight institutes covering the scientific areas of nanoscience and nanotechnology, informatics and telecommunications, materials science, chemistry and biology to nuclear physics and nuclear technology and radiopharmaceuticals (http://www.demokritos.gr). The objectives of LTP are to promote and diffuse scientific and technological work and achievements, to commercialize R&D results, to offer specialized services to the private and public sectors, and to contribute to the development – through technological innovation – of the knowledge-based society. The incubator of LTP hosts companies activating in the fields of biotechnology, informatics, materials, energy and services.

In the same period with LTP, another science park emerged in a city near Athens, Lavrio, under different terms, the Lavrion Technological and Cultural Park (LTCP). It was founded by the National Technical University of Athens, with the collaboration of local institutions, the people of Lavrion and the support of the Greek State and the European Union. The history of the area started in 1876, when "The Companie Francaise des Mines de Laurion" developed the mines of Lavrion. Lavrion was the first city in the newly created state of Greece to come across a great economic, industrial and cultural development. In 1989, the differentiation of employment in Europe had a harsh effect on Lavrion leading to the closing down of all industries one after the other. The intervention of the Greek state to save the mines was not capable of avoiding the catastrophe. After more than a century the mines were shut down forever leaving a whole city without employment. In 1992, the Greek government, in a movement of encouraging local development in Lavrio, bought the whole area of the old industry and handed it over to the National Technical University of Athens with the aim of creating a new pole of attraction in high technology and culture. The foundation of the Lavrion technological culture created and functioned soon after its restoration (http://www.lavrioferenceculturalpark.gr).

One year later, in 1993, another science park emerged in southern Greece in Crete. The Science and Technology Park of Crete (STEP-C) and the Managing Company of STEP-C (EDAP SA) were established in 1993 in Crete. The idea for a technology park in southern Greece, especially in Crete, dates back to 1988, when it was first designed by key individuals in the Foundation for Research and Technology-Hellas (FORTH), one of the most respectable research institutes in the country.

STEP-C aims to provide to the significant research activities of FORTH with a reliable interface to the business world and to have a significant role for the development of the region. Another purpose is to support company members of the STEP-C to exploit any novel technology opportunities offered by the research institutes and to become key vehicles in the technology transfer process. Its main purposes are the exploitation of inquiring results and the creation of a pole of growth in the region, besides the poles of the primary sector and tourism, the support of installation and growth of new enterprises of high technology and the creation of a "Centre of Learning" for executives of enterprises. The various services the STEP-C covers the needs of the enterprises, not only those hosted in its incubator but also the companies situated in the wider region. Those services are academic, technological, entrepreneurial, financial, legal, and informative (http://www.stepc.gr).

Thessaloniki Technology Park (TTP) was established in April 1995, by the Chemical Process Engineering Research Institute (CPERI). Its purpose was the greater exchange of ideas, people and facilities between universities and industry, especially those in northern Greece. The TTP building infrastructure has a total surface of 7,500 m^2, including the Centre for Research and Technology Hellas (CE.R.T.H.)/CPERI research laboratories/pilot plans, an incubator building and an administration/conference centre and library/scientific information. The incubator is available to enterprises and to individuals who want to convert an innovative idea to an enterprising success. Incubator services include accountant, secretarial support, connection with Internet and ISDN, e-mail, searches of collaborations and support on attendance in European and national programs E&A. (http://www.thestep.gr).

In the Region of Epirus, the Scientific and Technological Park of Epirus (STEP-EPIRUS) was founded in 1999 from the University of Ioannina and the General Secretary of the Region of Epirus. The main purpose of STEP-EPIRUS is the diffusion of the know-how that is produced in the academic community and the research centres aiming at creating a new pole of development for Epirus. It serves as an *incubator* for new enterprises among the four prefectures of the Region of Epirus. The enterprises hosted on the premises of STEP-EPIRUS develop high-technology products and services. In parallel, STEP-EPIRUS offers to the hosted enterprises consultant services, and is a mediator to academic and research institutions, collaborations with other enterprises, administrative and secretarial support (http://www.step-epirus.gr).

The incentives for the second science park to be settled in central Greece were given by the private sector in the Region of Thessaly, in Volos. The Technology Park of Thessaly (TE.PA.THE.) was established in December 2001 by the Metallurgical Industrial Research and Technology Centre S.A. (MIRTEC) and 38 other shareholders, mainly agencies and companies of the region of Thessaly. TEPATHE is a new model of collaboration including industrial, academic, research and government organizations, and was established in order to lead the knowledge-based information society of the twenty-first century in the Region of Thessaly. Its main objectives are the acceleration of the establishment of new dynamic high-technology companies, the encouragement of the improvement of existing companies with the introduction of new technologies, and the support of local and regional development.

The Technology Park of Thessaly promotes activities that contribute to the increased competitiveness of the Thessalian Industry. This is achieved by participation in many European and National Regional Development programmes. In TEPATHE, the Technology Transfer Unit contributes to the transfer of research and other activities products of research institutes and universities to the regional industry. The Technology Park of Thessaly has as a short-term development strategy to start out as an incubator for small firms originating from regional higher education institutes, research centres and the local community in the Region of Thessaly. In addition, the contribution of TEPATHE in the development of the Region of Thessaly is of great importance (http://www.tepathe.gr).

According to the frame presented above, there is a differentiation among the objectives and the legal framework of science parks in Greece. Their aims, the institutions of their administration and collaboration (universities, local councils, private organizations, etc.) appear to be adapting depending on the particular characteristics of the region of installation of each science park. However, an institution that could meet the needs of all the science parks in Greece, promote the exchange of information of technology and extend considerably the market of companies that support and engrave common results, is the Hellenic Science and Technology Parks Association (HESTEPA), founded in July 2006.

HESTEPA aims at growing narrower collaborations between existing but also future Greek organizations. In particular, its goals are:

- The facilitation of communication between its members
- The formulation of proposals for the appointment and promotion of national policy with regard to the institution of science and technology parks

- The promotion of the role of science and technology parks in the local and regional growth
- The creation of networks of collaboration with other institutions and particularly with the enterprises that are hosted in science and technology parks as well as with institutions of enterprises
- The intensification of mechanisms of diffusion of technology with the collaboration of science and technology parks, including academic institutions, government-owned institutions and enterprises
- The participation in international organizations with relevant goals
- The attendance in national or international committees, councils, conferences, reports, etc., related to the aims of HESTEPA
- The organization, attendance and concretization of programs of training and education
- The collection, organization and diffusion of information relative to the interests of its members
- The organization of meetings, congresses and the publication of special forms that promote the objectives of its members
- The provision to other organizations of services in subjects that are related with the objectives of its members
- The attendance in programs and actions that support the objectives of its members

Although science and technology parks (STEPAs) in Greece have been extendedly promoted, their status needs to be further developed. Some general directions concern with the exploitation of HESTEPA, so that new synergies will be created. In addition, the Greek universities and research centres, and especially their liaison offices, should be advanced and gradually linked with the Greek science parks (Bezitzoglou 2006). Also, all Greek STEPAs should be supervised by the same governmental organization, so that strong bonds will be developed between the two parties. The new role of science parks may be to cater to the development of the social capital necessary for enabling and facilitating entrepreneurship in networks (Hansson et al. 2005). However, since Greece does not belong to the core technological countries of Europe (Bakouros et al. 2002), above all and the most important is the development of a strategic plan for innovation, which will provide the financial and legal framework and the general terms for successful incubator services and support for enterprises in Greece.

2.4 Outsourcing Networks in Ireland

C. Heavey, P. Liston and P. J. Byrne – University of Limerick

Ireland experienced a disproportional level of foreign direct investment in the 1990s resulting in the manufacturing sector being dominated by large multi-national companies. These multi-nationals were attracted to Ireland by a combination of the competitive corporation tax system introduced by the Irish government in the 1980s, relatively low labour rates (although this has steadily increased since then) and the emergence of Ireland as a knowledge economy. Consequently, Irish SMEs typically participate in virtual networks through outsourcing contracts with these multi-national companies, or as they are sometimes termed OEMs (original equipment manufacturers). These OEM-lead outsourcing networks are a defining characteristic of modern Irish manufacturing.

Outsourcing networks (or virtual networks) typically link highly innovative but de-verticalized lead firms (OEMs) with sets of highly functional suppliers who provide a wide range of production-related services (Sturgeon 2000). These networks are highly flexible systems characterized by fluid relationships with short-term contracts between participants within the network. Such networks are now a characteristic of our business nation and can be found extensively across the Irish landscape. These networks are of extreme importance to the emergence, survival and growth prospects of many of our indigenous SMEs. In recent times the impact of our knowledge economy has surpassed the competitive edge we once enjoyed with respect to our low labour rates. SMEs have had to adapt to these changing conditions. In the past SMEs competed through the provision of low cost manufacturing to support OEMs, they now compete through superior supply (network) management for these OEMs. So in summary, indigenous SMEs, working with these OEMs have moved through an evolutionary phase from low-cost manufacturers to their present day position of knowledge-based network managers.

Section 2.4.1 reports on a field study carried out in current practices in outsourcing with particular focus on the role of networks of SMEs. This study found that the formation of Irish outsourcing networks is typically instigated and aided by what is known as the RFx process (where RFx is the collective term for *request for information*, *request for proposal* and *request for quotation*). This RFx process and its implications for the responding companies are discussed in Sect. 2.4.2. Sections 2.4.3, 2.4.4, and 2.4.5 then describe three different network structures based on case study examples of indigenous Irish companies. These case studies highlight the varying levels of interaction and forms of governance that exist in Irish manufacturing networks today.

2.4.1 Electronics Manufacturing Field Study

This study focused on the electronics sector as it is the largest manufacturing industry in Ireland with over 30,000 people involved in a wide range of sub-sectors including computer systems and sub-systems, peripherals and media, electronic components, data communication equipment, control and test systems and consumer electronics (Shannon Development 2007). The supply chains involved with producing electronic goods invariably contain many echelons with various com-

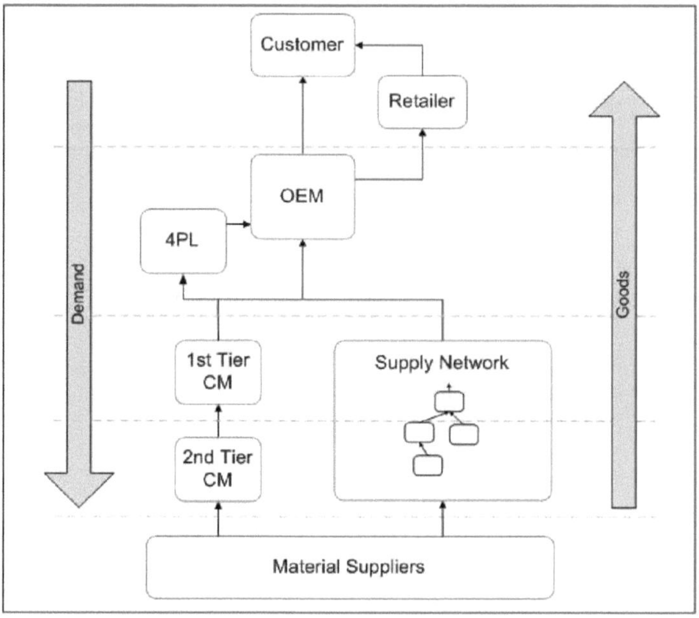

Fig. 2.7 Relative company positioning in typical supply chain networks

panies playing different roles in the production process (due primarily to repeated outsourcing activity), as depicted in Fig. 2.7.

In order to gain a balanced understanding of current outsourcing practice, field study participants were selected from across the supply chain spectrum (the supply chain roles of the participants are listed in Table 2.3). In all, seven different organizations are included in this field study evaluation, varying in size from a network coordinator with only four direct employees to a large multi-national OEM with over 50,000 employees. One of the organizations included in the field study is an association of over 40 individual companies and is best described as a virtual breeding environment (VBE). A VBE is defined by Afsarmanesh and Camarinha-Matos (2005) as "an association of organizations (members) and their related supporting institutions, adhering to a base long-term cooperation, agreement, and adoption of common operating principles and infrastructure, with the main goal of increasing their preparedness towards collaboration in potential virtual organizations". Such a virtual organization (VO) (also referred to as a virtual enterprise (VE)) is defined as "a temporary alliance of enterprises that come together to share their skills, core competencies, and resources in order to better respond to business opportunities, and whose cooperation is supported by computer networks" (Camarinha-Matos 2001).

During the course of the field study it became apparent that it was not only the case of the VBE which resulted in the formation of VOs. It was noted that in most cases the companies did not undertake the outsourced business in its entirety but instead formed an alliance with other companies which they relied upon for

Table 2.3 Field study participant details

Field study participant	Supply chain role	Company's main area of business	Number of employees in company (approx.)	Number of RFQs processed in a week (approx.)
Participant one	2nd Tier CM	Low volume assembly and manufacturing	130	5
Participant two	1st Tier CM	Electronics manufacturing	50	6–12
Participant three	1st Tier CM	Electronics manufacturing	600 (Irish facility)	8
Participant four	4PL	Warehousing, kitting and supply chain management	400	5–10
Participant five	Supply network coordinator	Supply chain management	4	4–5
Participant six	Supply network	Manufacturing related services	46 Independent companies	Still in initial stages of business development
Participant seven	OEM	Electronics manufacturing	>50,000 (globally)	n/a

materials and services. This leads to the proposition that there exists a VO continuum which spans various levels of complexity and integration in the involved networks.

Based on the field research, three different VO structures from this continuum are outlined in Sects. 2.4.3, 2.4.4 and 2.4.5. Each of these descriptions is supported by an Irish example from the study which typifies the type of network concerned. First, to give the context in which these supply networks are formed, the RFx process is described in Sect. 2.4.2.

2.4.2 Network Creation: The RFx Process

This section presents the RFx process which is encompassed in many different terms in both industry and literature (e.g. supplier sourcing, contract costing, supplier selection, partner search and selection, and tendering). The RFx process is of significance as it instigates the creation of VOs in order to meet the requirements of an RFQ. Having made the decision to outsource a business process the company (be they OEM or contract manufacturer (CM)) will first of all identify a broad selection of potentially suited contractors. This group of potential contractors is then reduced in size by subjecting it to a number of iterative supplier analysis steps (see Fig. 2.8).

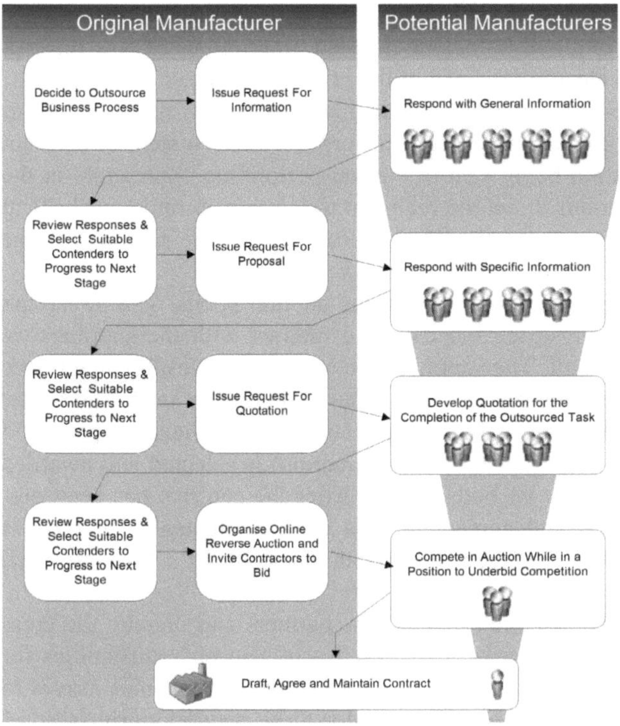

Fig. 2.8 The RFx process

This description is typical of the outsourcing decision process when the company outsourcing the business (described here as the buyer) is unfamiliar with many of the potential suppliers' capabilities. However, depending on the extent of the buyer's knowledge of the suppliers in review, they may choose to forego some of the steps identified here.

Once the potentially suitable contractors have been identified, the buyer issues what is known as a request for information (RFI) to each company in the selection. This request is to simply elicit general information about the companies and their competencies. Responses to an RFI are not binding on the respondent and are used more to eliminate non-relevant and non-responding contractors rather than select between vying service providers. Once the relevant contractors have been identified they each receive a request for proposal (RFP), which outlines the business process to be outsourced and requires a plan from the contractor detailing the system they propose to put in place in order to undertake this business process. The contractors are not required to provide cost information at this point in the process as future contenders are selected on the basis of the performance and capability they purport to offer. The contractors who have been successful in these first two steps of the process are now presented with a request for quotation (RFQ). This RFQ document contains a precise description of the work to be completed and outlines all other customer requirements and contractual agreements. Responses to an RFQ are required to include detailed descriptions of how the task will be undertaken, with a clear breakdown of the pricing scheme.

The successful contractor is often chosen at this stage but in recent years reverse auctions have been increasingly used as a further selection step. These auctions begin with an invitation being sent to the top performing contractors in the RFQ process. These invitations direct the recipient to log into an online website at a specified time where they are required to underbid (hence the phrase "reverse auction") each other in an attempt to win the contract.

The suppliers must prepare their bid before the auction opens. The buyer can watch over the process, analyse the bid data and interact with the suppliers by phone, message board or e-mail. The suppliers see only the "lowest" bid and have to beat it to win the auction. This "lowest" bid is based on criteria set by the buyer, and can contain multiple attributes. When the auction has finished, the preferred supplier(s) (optionally the winner of the reverse auction) is selected and awarded the contract, pending agreement by both parties. After the contract has been prepared, drawn up, negotiated and agreed upon, a contract management process begins where the contract is managed throughout the contract's lifecycle until termination of the contract.

This process supports the selection of business partners and thereby the creation and extension of supply networks. However it can also pose difficulties for networks, particularly in cases where an existing network of companies strives to collectively respond to an RFQ. The difficulties for these networks primarily relate to the short timeframes in which they must gather the required information from individual members and form a united solution for the buyer. The management structure of the network can also be a major factor when decisions have to be made quickly during a reverse auction; three different structures are discussed in the following sections.

2.4.3 Case 1: Contract Manufacturer

A contract manufacturer (CM) can be defined as follows (Bridgefield 2006): "A third party that performs one or more production operations for a manufacturer who will market the final item under their own name. They often charge on a per-piece or per-lot basis for the labour required for their services while using components or materials supplied and owned by the final item manufacturer." Contract manufacturers can be further categorized depending on their positioning within a supply chain. The terms *first tier*, *second tier*, etc., are used to denote these positions. When responding to an RFQ, a CM will very often be required to develop a supply network of component and assembly suppliers. While the CM may not develop formalized associations with such companies, they have to form a close enough relationship to ensure they have a secure supply of material. This is particu-

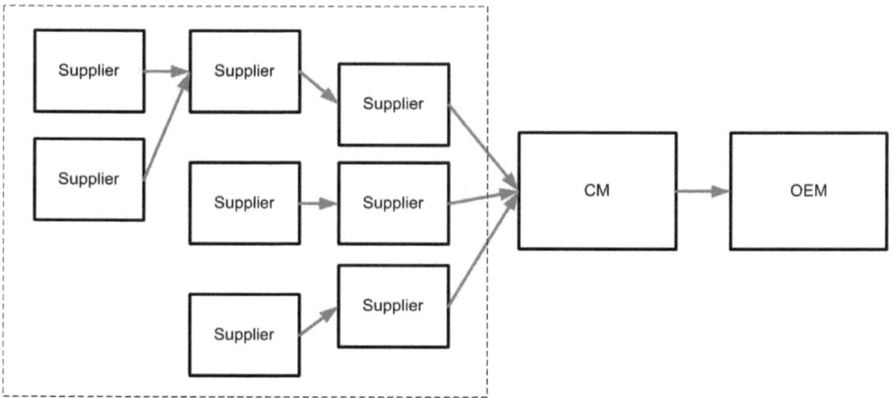

Fig. 2.9 CM supply network scenario

larly true in situations concerning specialized commodities that are not widely available on the open market or those for which the CM is the supplier's sole customer.

Many of the RFQs examined during the field study required the CM to develop one of these informal networks. One example concerned an OEM that had decided to outsource all of its packaging activities. To bid for this work, the CM was required to develop and cost a solution where it would be responsible for managing all material, procurement, warehousing, inventory, IT, logistics and quality activities. This work involved relatively straightforward processing on the part of the CM (i.e. packaging and labelling of goods). The key consideration for the CM was therefore ensuring that all the necessary packaging materials were available at all times as any delay in the shipping of goods would incur severe financial penalties from the OEM. To achieve this, the supply network as illustrated in Fig. 2.9 was developed.

The CM in this case had six weeks to respond to the RFQ. In this time they had to secure suppliers for moulded plastic cartons, assorted corrugated cardboard boxes, and various labels; agree stock holding levels at each supplier; and design contingency plans for critical materials. Costs for providing this service had to be calculated based on supplier quotes, estimated expediting costs, CM labour rates and overheads. The final decision on the contract winner in this example was based on the result of a reverse auction.

2.4.4 Case 2: Supplier Sourcing Company

A supplier sourcing company is an organization that selects suitable organizations on the open-market to fulfil the needs of an OEM. During the field study it was noted that many of the companies operating in this role in Ireland were former manufacturers that changed business focus. These companies use their manufacturing experience to their advantage by distinguishing potential processing problems

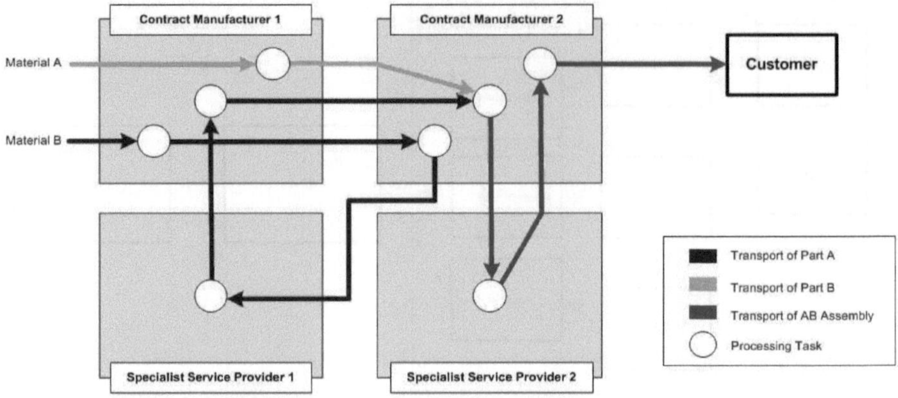

Fig. 2.10 Process steps for component supplied by a supplier sourcing company

and using their industrial contacts to source service providers with the expertise required. The supplier sourcing companies then take on responsibility for managing the flow of material between these independent companies and on to the customer.

One common reason why a supply network is needed in these circumstances is that there simply may not be a single company with all the technological capabilities required. A second reason is that, even when one company is capable of completing all tasks, there may be other companies who can complete specific tasks more efficiently and/or more economically.

Participant five from the field study is a typical example of a supplier sourcing company. Figure 2.10 illustrates a supply network designed by this company to supply a highly specialized product to a multi-national OEM. Four different companies based in two different countries are involved in the manufacturing process; therefore, significant transportation of the components is required. This necessity to use foreign specialist partners poses a greater problem for Irish companies than their mainland Europe counterparts as any cross-border collaboration requires either sea or air transportation. Note from Fig. 2.10 that product components visit both Company 1 and Company 2 at three different times during the production process, thereby further increasing logistical and scheduling problems for the network manager.

2.4.5 *Case 3: Virtual Breeding Environment Supply Network*

The third identified supply network structure is typified by *participant six*. This network is an open alliance of companies in the Shannon region of Ireland that was established in January 1999 and is best described as a regional VBE.

The network was originally established to generate more business opportunities for the member companies through joint marketing initiatives such as exhibitions, tradeshows, and advertising. However in addition to this *participant six* is now

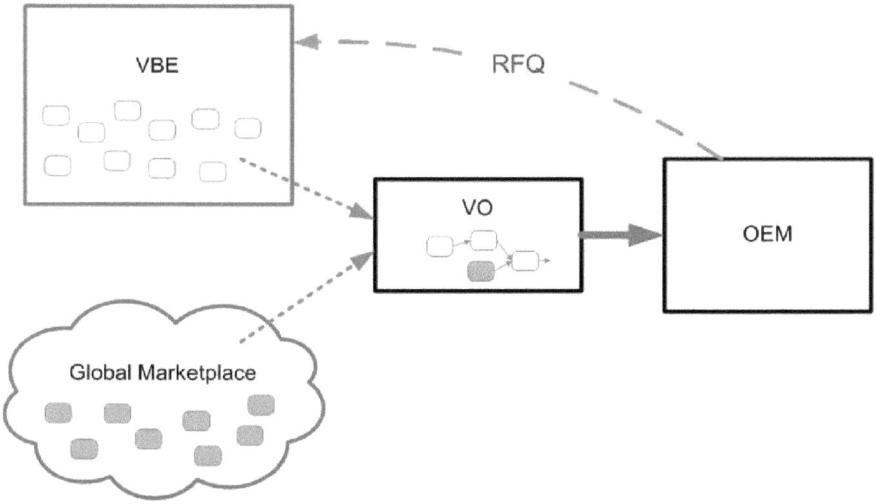

Fig. 2.11 VBE supply network scenario

aiming to increase the capability of the network members to collaborate when responding to RFQs so that they may rival larger competitors. An overview of the creation and operation of the network is illustrated in Fig. 2.11. When an RFQ is sent by an OEM to the VBE (this may be directly to the governing board of the VBE or to one of the individual members who in turn brings the business opportunity to the attention of the entire VBE), a network member is elected as VO developer and thereby made responsible for creation, formation and management of the network. The RFQs handled by *Participant six* are typically multi-faceted, requiring the VBE to consider the design, manufacture and product lifecycle management of a family of products. The VO developer must align each required task with a suitably skilled network member which, as illustrated in Fig. 2.11, may require the participation of companies from outside the VBE.

With the experience of operating in this manner, *participant six* has identified many barriers that impede the development of VO networks, particularly those with a high number of required members. The most significant constraint is reported to be the shortage of explicit supply network design tools and methodologies for developing, costing and selling a VO supply network solution to the customer.

2.4.6 Summary

Recent changes in global economics have lead to the transformation of business practice for many Irish SMEs. Where companies were once able to operate independently and rely on low labour rates for competitiveness, they now rely on collaboration and quality of service to ensure their continued operation. This has promoted the development of formal and informal outsourcing networks which span both across the island of Ireland and into many other countries; three different network structure examples have been outlined here. As SMEs begin to either participate in or manage these networks they require new collaboration support systems; it is important that these requirements are addressed in order to secure the future of Irish manufacturing SMEs.

2.5 Industrial Districts in Italy

M. Salvador and S. Salvador – EIDON

Industrial districts are established as an important asset of the Italian production system: the numbers of active companies, employees and exported products make them an important part of the national economy. Italian industrial districts are generally recognized as a valid organizational model, studied both on a national and international level, which makes member companies and districts as a whole capable of competing on international markets.

The international interest in SME clusters has been fuelled particularly by the experience of what has come to be called the Third Italy. The concept of the Third Italy started to be used in the late 1970s. At that time, it became apparent that while little economic progress was in sight in the poor south (Second Italy), the traditionally rich northwest (First Italy) was facing a deep crisis. In contrast, the northeast and centre of Italy showed fast growth, which attracted the attention of social scientists. In a number of sectors where small firms predominated, groups of firms clustered together in specific regions seemed to be able to grow rapidly, develop niches in export markets and offer new employment opportunities.

According to economical analysts "the growth of the northeast and centre of Italy pushed the scholars to analyse the economic and social fabric of the region, and its agglomeration of firms clustered in specific geographic zones, and operating in specific industrial sectors. These clusters were able to establish strong positions in world markets in a number of traditional product categories, including shoes, furniture, tiles, musical instruments, etc. Progress seemed promoted by the capacity of the clusters to innovate in terms of production processes as well as product qualities" (Callegati and Grandi 2005).

2.5.1 Basic Features and Figures of Industrial Districts

The Italian law no. 317, 5th October 1991, which rules state support for innovation and development of SMEs, defines industrial districts as: "geographical areas characterized by a high concentration of Small and Medium Enterprises, with peculiar reference to the enterprises' connection with locally resident population and productive specialization of the assembly of SMEs as a whole". Furthermore, it states that (Districts') Consortia for Industrial Development are to be considered public economic bodies. It then legitimates the regions to financially support innovative projects that involve many enterprises, according to specific contracts that are to be stipulated between the regional governments and the consortia. Priorities for financial intervention are decided by the regions.

The reference *Guide to Italian Districts* published by Unioncamere identified 100 existing districts in the country, gathering some 89,000 enterprises, which in turn employ almost a million workers (see Table 2.4).

From this common starting point, Italian regions have operated in differing ways, thus creating an inhomogeneous national landscape with regards to industrial districts' spreading and importance. The following regions, Piemonte (29 districts), Lombardia (16 districts) and Veneto (29 districts), gather alone more than 50% of all Italian districts (figures refer to 2004). Other regions, such as for instance Emilia Romagna, have preferred not to recognise districts as institutional entities until recent years, even though they carried out specific support actions for the nationally important local "clusterings". Friuli Venezia Giulia's regional government recognized four different districts with a regional law of 1999 and defined some specific support programs that will be analysed more in depth in the following paragraph.

Both the national and regional laws adopt a definition that recalls only two of the various characteristics on which the scholarly definition of "industrial district" is based. The points recalled in both laws are (1) a territorially restricted area and (2) a specific product or sector.

A more exhaustive analysis of the concept of "districts" (and "clusters") goes beyond the idea of locally defined areas of productive specialization (Becattini 2003).

The fundamental configuration of an industrial district, on a deeper level, implies that the production process is not integrated on a vertical basis but instead is based

Table 2.4 Italian industrial districts (adapted from Club dei Distretti Industriali e Unioncamere, Guida ai Distretti Industriali Italiani, Roma, 2003)

Sector	Number	Companies	Employees
Food and agriculture	8	4,072	59,317
Textile: apparel	19	24,175	225,413
Shoes	14	10,889	105,744
Mechanical	7	7,041	92,742
House appliances	13	14,548	129,300
Non-metallic minerals	14	7,128	57,305
Other sectors	25	21,010	326,331
Total	100	88,863	996,160

on work distribution between different companies which are inter-related within a single *supply chain.* The specialization that defines the territory, according to the legal definitions, applies to the end products (shoes, chairs, glasses, tiles, etc.) but it implies a further spectrum of process-specific specializations. Companies forming a district (cluster) are not only geographically close to each other but are necessarily inter-dependant and cannot be considered as completely autonomous entities.

Vertical cooperation often coexists along with intense horizontal competition between the districts' actors. Furthermore, not only direct production connections are active within a district: local institutions, industry and trade associations, banks, research and educational structures can be considered among the prime actors in the building and operation of a district.

Industrial districts in Italy emerged as a rather "spontaneous" way of organizing production for competitiveness but the role of local institutions played a not marginal role by means of relevant social regulation and public services for the enterprises.

Over time districts have been varying the range of products, both enlarging the products' spectrum within the original specialization field (e.g. from wooden chairs only, to metal and plastic chairs as well) or even trespassing the borders with neighbouring fields. *Product mix* enlargement derives from two opposite business strategies whose interaction produces a common trend for the district as a whole: product range expansions for some companies and product specialization for others, usually smaller ones.

The birth of new specialization has been parallel to the evolution of the inter-sectorial configuration that sees local companies broadening their expertise to the production of district-relevant machinery, materials and connected technologies.
Following this evolutionary scheme the districts, even though retaining their original nature, grow into more complex aggregations that correspond to Michael Porter's definition of *geographical clusters* (Porter 1998a):

"Clusters encompass an array of linked industries and other entities important to competition. They include, for example, suppliers of specialized inputs such as components, machinery, and services, and providers of specialized infrastructure. Clusters also often extend downstream to channels and customers and laterally to manufacturers of complementary products and to companies in industries related by skills, technologies, or common inputs. Finally, many clusters include governmental and other institutions – such as universities, standard-setting agencies, think tanks, vocational training providers, and trade associations – that provide specialized training, education, information, research, and technical support."

Districts have traditionally operated as closed systems, with a dense network of internal connections but a rather limited number of external interactions and have mostly relied upon the capability of generating internally the required human, financial and knowledge resources. Partly due to the evolution of the district model itself and mostly due to the unavoidable change factor of globalization, this traditional closed model is, although, undergoing a deep crisis and districts are confronted with the need for major changes. According to the cited *Guide to Italian Industrial Districts* it is predictable that industrial districts will not be able to face "big armies (multi-national companies)" without transforming themselves into "pocket multi-national companies" and acquiring a cosmopolitan dimension (Club dei Distretti Industriali e Unioncamere 2003).

De-localization of the production and distribution processes, development of strategic relationships with extra-district suppliers (especially with regards to services such as technological innovation, design, marketing and financial services): these are the factors that have characterized the last few years for most of the Italian districts.

The internationalization of the districts opens up new challenges and problems. Districts are faced with the task of changing from closed local systems to specialized international cluster networks, while at the same time retaining the characteristics that made districts successful in the first place and meanwhile avoiding loss of competitiveness in a an increasingly rough global market.

Knowledge-intensive business services (KIBS), such as the examples mentioned above, are becoming increasingly needed in all production fields and even more so for industrial districts that need to grow their contacts and connections to international networks.

The CODESNET project highlighted some of these challenges and provided some case studies of innovative high-tech and knowledge-intensive clusters that are growing on a European level and having their origin in Friuli Venezia Giulia.

2.5.1.1 A Remark Concerning Terminology

The term "industrial clusters", rather than "industrial districts" or "*milieu innovateur*", spread in the 1990s, a result of Porter's work, *Competitive Advantage* (Porter 1985) and his further publications. According to Porter's definition "Clusters are a geographically proximate group of interconnected companies and associated institutions in a particular field linked by commonalities and complementarities. Clusters encompass an array of linked industries and other entities important to competition … including governmental and other institutions – such as universities, standard setting agencies, think tanks, vocational training providers and trade associations" (Porter 1998b).

For the purposes of this chapter we will stick to the different terminologies currently in use, often using both the term "cluster" and "district" in a very similar meaning.

2.5.2 SMEs Clusters and Institutional Support: the FVG Case Study

Friuli Venezia Giulia (FVG), Veneto and Trentino Alto-Adige, are the regions where the "Italian northeast" economic model first emerged in the 1970s and further evolved becoming highly successful and studied on a national and international basis. The main characteristics of the model are a capillary presence of enterprises on the geographical area, a dynamic productive structure highly inclined towards export and innovation, a balanced mixture of craftsman businesses, SMEs and a certain number of larger companies (actors).

Among the main peculiarities of the northeast there is the presence of industrial districts, areas where a large number of companies operating in different production are gathered and whose production is actively integrated.

In the region Friuli Venezia Giulia four of such districts exist and have for many years been well-consolidated. They are the Chair district around Manzano, the Furniture district in the Pordenone area (Brugnera), the Knife district in Maniago and a Quality Food district, especially focused on specialized ham production, in San Daniele[2] (a detailed description of their main characteristics can be found on the CODESNET website, as shown in Chap. 5).

Next to these established districts, new and innovative associations and clusters are emerging, tied to new technologies and advanced services: ICT, nanobiotechnologies, naval mechanical. In all these knowledge-intensive fields Friuli Venezia Giulia aims at better exploiting the region's quite distinctive position, both from a geographical point of view (its location in the centre of the new enlarged Europe) and from the perspective of number and quality of the territory's human capital, when compared to the other Italian regions.

A high level of educational resources (see statistics on the number of R&D employees per inhabitants as well as figures about number of scientific publication and their impact factor; see Fig. 2.12) contributes to helping the economical dynamism of the region, and investments in research and innovation are ever more needed because of the challenges and crisis following the increased scale and velocity of global competition and the emergence of competitors on the international market.

Relevant attention has been attributed by the institutions to supporting growth and activity of SME clusters and industrial districts. In the Regional Law no. 27, 11th November 1999, the region formally recognized industrial districts as "contexts of economic and occupational development and seat for promotion and coordination of local initiatives of industrial politics (…) to the purpose of reinforcing competitiveness of the productive system, the efficacy of the existing tools for industrial politics and of defining and activating new policies and guidelines". The same law defines the composition of a governing District Committee (Consortium), financed on a mixed public-private basis, which should include local authorities' representatives, representatives of the Industrial and Craftsman Associations and labour representatives. For the aims of supporting the activity of the districts, the District Committee produces a three-year development program plan, which is the basis for financial negotiation with the regional government.

---

[2] Respectively named Distretto della Sedia (PROMOSEDIA), Distretto del Mobile Livenza (DML), Consorzio Coltellinai Maniago, Distretto alimentare di San Daniele.

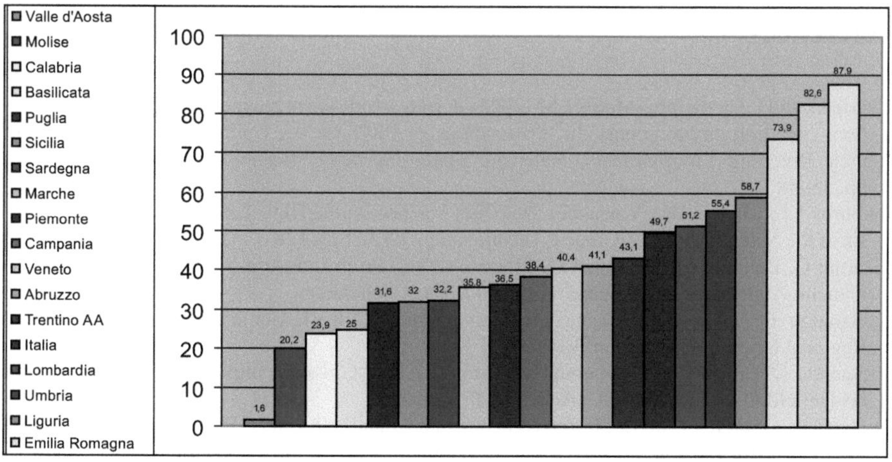

Fig. 2.12 Number of scientific publications per 100,000 inhabitants, Italy 2005 (adapted from IRES-FVG on data from ISI Web Science and ISTAT)

In more recent years the regional government, partly as an answer to new threats to the traditional economy of the territory and its traditional markets, has further expanded the set of institutional tools for supporting SMEs and SME clusters.

Recognizing the importance of innovation and research for regional enterprises as means of development and of maintaining a competitive advantage over global competitors, the regional government has resorted to some focused intervention for building or supporting the structuring of a certain number of new high-tech clusters. At the same time, it planned and executed some specific action to help existing clusters renew their competencies and push investments on innovation and knowledge-based development. Another Regional Law, no. 4 of 4th March 2005, "Interventions for support and competitive development of SMEs" ensures regional financial support to companies proposing well-structured plans for dimensional and competitive growth.

The institutional support activities helped shape the new cluster for home and living technology ("FVG Abitare" district), the one for biomolecular medicine in Trieste and to redesign the cluster of naval/electromechanical industry in Monfalcone. The region also sketched specific strategy plans for development of ICT and biotech/biomedicine fields, in both cases including points to support new cluster-like concentration of competencies and entrepreneurship.

Finally, and promisingly for the future, thanks to the enlargement of the European Union and thanks to the growth of the emergent countries in Southeast Asia, Friuli Venezia Giulia finds itself again in a central position. Its strategic position has stimulated a strong thrust towards innovation and know-how transfer from laboratories to enterprises.

References

Afsarmanesh H, Camarinha-Matos LM (2005) A framework for management of virtual organizations breeding environments. In: Proceedings of PRO-VE'05, Collaborative Networks and their Breeding Environments, Valencia, Spain. Springer, Berlin Heidelberg New York, pp. 35–48

Bakouros YL, Mardas DC, Varsakelis N (2002) Science Parks, High Tech Fantasies? An Analysis of the Science Parks of Greece. Technovation 22(2):123–128

Becattini G, Bellandi M, Dei Ottati G, Sforzi F (2003) From Industrial Districts to Local Development. An Itinerary of Research. Edward Elgar, Cheltenham

Bezitzoglou C (2006) Showcasing Innovative Greece. ERSA Conference Papers. European Regional Science Association, Vienna

Bridgefield (2006) Bridgefield Group ERP/Supply Chain Glossary. http://bridgefieldgroup.com/bridgefieldgroup/glos2.htm#C. Accessed 17 Dec 2007

Callegati E, Grandi S (2005) Cluster Dynamics and Innovation in SMEs: the Role of Culture, International Centre for Research on the Economics of Culture, Institutions, and Creativity (EBLA), Working paper No. 03/2005. Dipartimento di Economia "S. Cognetti de Martiis", University of Torino

Camarinha-Matos LM (2001) Execution system for distributed business processes in a virtual enterprise. Future Generation Comput Syst 17(8):1009–1021

Chan KF, Lau T (2005) Assessing technology incubator programs in the science park: the good, the bad and the ugly. Technovation 25(10):1215–1228

Club dei Distretti Industriali e Unioncamere (2003) Guida ai Distretti Industriali Italiani, Roma

Dettwiller P, Lindelof P, Lofsten H (2006) Utility of location: A comparative survey between small new technology- based firms located on and off Science Parks- Implications for facilities management. Technovation 26(4):506–517

Hall P (1994) Technopoles of the World: The Making of the 21st Century Industrial Complexes. Routledge, London

Hansson F, Husted K, Vestergaard J (2005) Second generation science parks: from structural holes jockeys to social capital catalysts of the knowledge society. Technovation 25(9):1039–1049

IASP (1996) The economics of Science Parks. International Association of Science Parks, Malaga

IASP (1998) Delivering Innovation. International Association of Science Parks, Malaga

Mowery DC, Rosenberg N (1989) Technology and the Pursuit of Economic Growth. Cambridge University Press, Cambridge

Podolny JM, Page K (1998) Network Forms of Organization. Annu Rev Sociol 24:57–76

Porter ME (1985) Competitive Advantage. Free Press, New York

Porter ME (1998a) Clusters and the new economics of competition. Harvard Business Review, Cambridge, MA

Porter ME (1998b) On Competition. Harvard Business School, Cambridge, MA

Shannon Development (2007) Engineering & Electronic Manufacturing at Shannon Free Zone. http://www.shannonireland.com/ShannonIrelandSectors/EngineeringElectronics/. Accessed 6 Dec 2007

Sturgeon TJ (2000) How Do We Define Value Chains and Production Networks, MIT IPC Globalization Working Paper 00-010

Thomson G (2003) Between Hierarchies & Markets: The Logic and Limits of Network Forms of Organisation. Oxford University Press, Oxford

Williams IF (2000) Policy for Inter-firm Networking and Clustering: A Practitioner's Perspective. http://www. clusters.org.nz. Accessed Sept 2007

Chapter 3
Promote Aggregation of SMEs: Suggestions and Actions

Abstract The discussion in Chap. 2 presents common characteristics of SME aggregations in different European countries, but also describes some causes of potential crisis. SME clusters and networks essentially are groups of weak industrial bodies – as small enterprises surely appear – which try to enforce their individual weakness by stipulating an agreement for reciprocal help and support. This is the real problem: not only how to define this agreement, but mainly for which actions and common decisions. In developing the CODESNET project, the partners from different European countries have focused a few main critical aspects which should be improved in existing SME networks: on one side, the necessity of a stronger coordination, to be applied either by a central committee or by some commonly accepted consortium management; on the other, the opportunity of having at their disposal efficient tools for improving the performance of clusters and supply chains, as well as the competence of the employed personnel. These critical aspects and proposals to approach them, in terms of new development projects and also of new training and research plans, are analysed in this chapter. The set of contributions has been selected in such a way to give an overview of the new initiatives in different European countries.

3.1 Strengthening SME Network Governance

A. Villa and D. Antonelli – Politecnico di Torino

Making an overview of the SME clusters and networks described previously in Chap. 2, it appears a variety of networked industrial bodies, which operate in largely different industrial sectors, include different types of enterprises and agencies. But in all these networks a common structure can be recognized: a "graph" of connections over which both items and information are flowing, and therefore make the internal SMEs as cooperative as possible.

The graph-based structure of connections among SMEs is evidence of two main problems facing any SME network: on one hand, the necessity of a production-oriented pattern of links, organized in such a way as to facilitate the management of production flows in the network; on the other, the necessity of associating to the interactions graph a proper organizational chart, to efficiently and effectively manage the SME network as a whole.

Indeed, the problem of having a robust management of interactions is the main concern of a SME network.

This section will address the problem of analysing the need of a network management able to make all component SMEs as cooperative as possible.

The starting point will focus on the preliminary requirement for an effective network management: to deal with a clear and simply manageable pattern of the SMEs links. This first analysis allows to catalogue into a few graph models, the typical patterns of links, which can be recognized by reviewing the presented SME clusters and networks in Europe.

After that and based on the graph representations, a common conceptual model of any SME network in terms of "demand and supply network of autonomous agents" (Villa and Cassarino 2005) can be stated. This is simply the SME networks' representation which can motivate, in the clearest possible way, the necessity of a management unit of the network as a whole.

3.1.1 A Short Catalogue of Graph Models of SME Networks

Industry networks can be profitably modelled by means of graphs. "A *graph* in this context refers to a collection of *vertices* or *nodes* and a collection of *edges* that connect pairs of vertices. A graph may be *undirected*, meaning that there is no distinction between the two vertices associated with each edge, or its edges may be *directed* from one vertex to another" (Wikipedia 2008). A directed graph is usually represented by means of an arrow pointing to the destination vertex.

In the representation of networks, the graphs have n vertices, a one-to-one correspondence with the enterprises of the cluster. Each node is linked to another one in the graph if a relationship between the two corresponding companies exists, in particular if the two enterprises share some kind of information or if there is a flow of goods or information between them.

The simplest and the first application of the graph to networks is the modelling of a supply chain. A supply chain is the system of organizations, activities and resources involved in moving a product or service from supplier to customer.

The simplest supply chain involves a firm, its supplier and its client. A longer supply chain takes into account also the indirect suppliers and clients, i.e. the supplier's supplier and the client's client. A graphical representation of a supply chain is represented in Fig. 3.1.

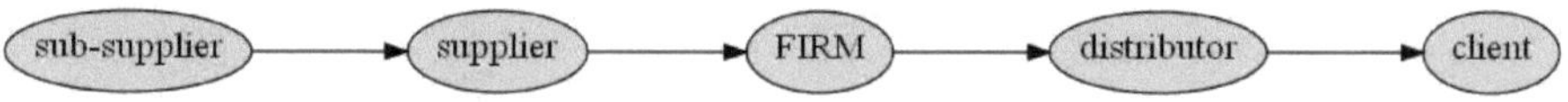

Fig. 3.1 Model of a supply chain

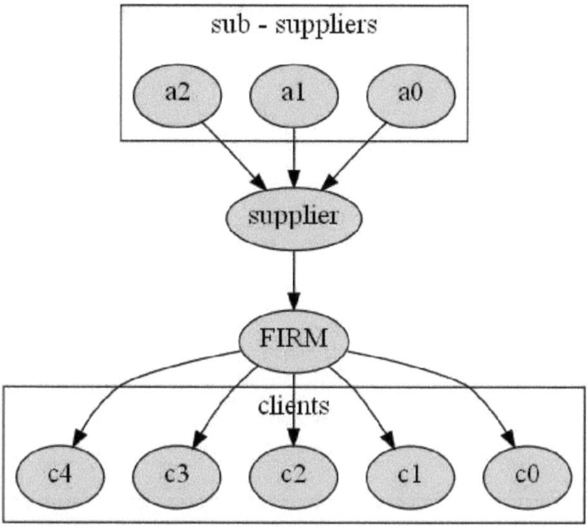

Fig. 3.2 Model of a multi-agent supply chain

It is apparent that this kind of representation is centered on one firm and considers all the others as external agents whose task can be only to supply raw materials or to buy finite products.

Supply chains can be further complicated without losing the firm-centered approach, as depicted in Fig. 3.2.

The obvious extension of the supply chain concept is to consider the possibility of multiple interactions among firms in a network, for example in the case of a joint venture in which more firms collaborate in the making of the final product without a clear distinction among suppliers and assembler.

A further kind of analysis refers to the peculiar role of each firm within its cluster. Following the conclusion derived from Becattini (1990) in his work based on statistical correlations, it is possible to identify two types of enterprises and, depending on their distribution, two types of network organization structures as well. Enterprises could be differentiated as primary and secondary according to their relative weight in their network of interactions.

Primary enterprises are characterized by a larger incoming or outgoing production flow within the network and with respect to all the others firms (these are secondary). A network could be composed of a single primary enterprise, as in Fig. 3.3a, and a set of secondary ones, or by a primary link, that is two (or possibly more than two) enterprises exchanging the larger percentage of the cluster flow. Figure 3.3b shows the example of a network composed of two primary enterprises,

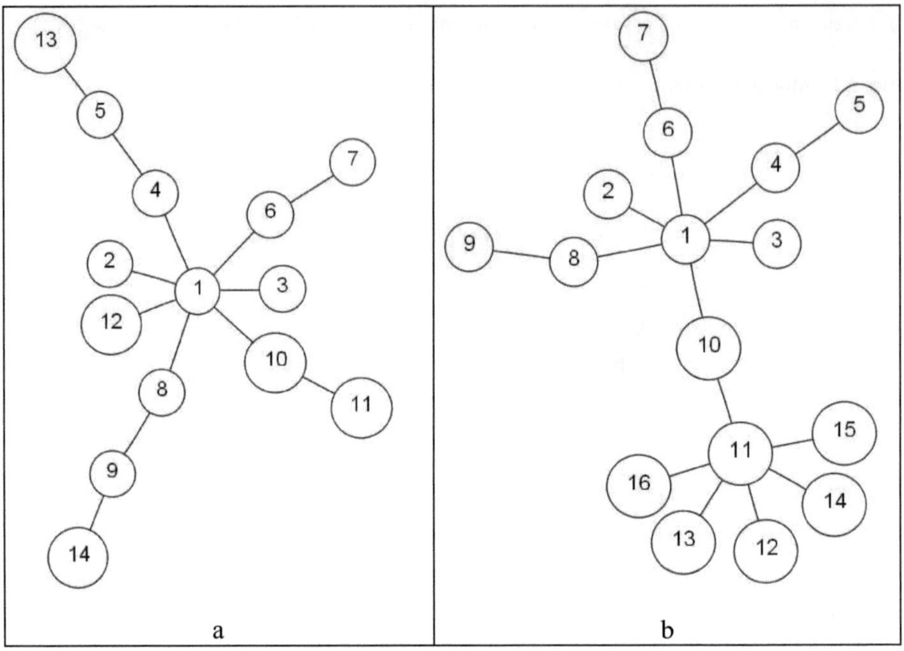

Fig. 3.3 **a** A "star" network with enterprise 1 playing the role of *primary* enterprise. **b** A *primary link* between two *primary* enterprises 1 and 11

named 1 and 11, each of them managing its own network of suppliers and connected by the common supplier 10. The primary enterprises have usually the bigger production capacity in the network and they probably also manage the link with other clusters, when they exist.

Some hypothesis about the networks organization structure could be derived. The organizational structure refers to the decision-making processes in the administration of the network (who makes decisions? who decides the volumes exchanged, the prices?).

When a single primary industry exists, the network has a hierarchical organization, while in the second case the structure is said to be polycentric. When, on the contrary, no significant difference can be recognized in the flows incoming and outgoing from firms belonging to the network, the network is similar to the one represented in Fig. 3.4 and is said to be canonical (the theoretical definitions of the three types of networks are from Bititci et al. 2004).

In a hierarchical network the primary firm plays the role of main coordinator and leader of production, distribution and innovation processes. Secondary firms work in a commitment and their existence depends on the primary industry's capability of gaining market share near to the end customer. The potential role of public institution in supporting this kind of network is very narrow, at least until the secondary enterprises stay in the leader's shadow.

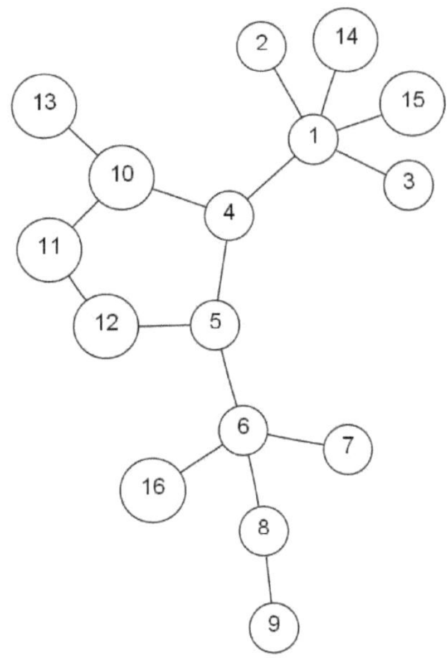

Fig. 3.4 A canonical network where it is not possible to distinguish enterprises, which are far more significant from the others

In a polycentric governance structure each primary industry is called to coordinate itself with both its sub-network of secondary partners and with the other primary firms. This type of organization typically characterizes products with a high degree of technological features, where the cognitive partition of the labour is crucial. The role of public organizations could be significant in assisting the primary firms in their role of multiple coordinators.

The canonical cluster refers to the typical network structure. It is composed of a network of demand and supply relationships centred on the production of the same type of end product. The network is nearly balanced, sufficiently open to the outside, highly socially and territorially characterized, and no strong formal relationships exist among firms. They are rather put together by habitual links strengthened by the physical and cultural closeness. Typically these networks produce high-quality products highly related to the resources (both material and not) available in the region. Public institutions have the crucial role of defending the tricky existence of both products and enterprises assisting them in critical matters of intellectual property, trademarks, marketing and innovation (assisting, for example, the creation of common laboratory of research or technological consulting).

The organizational structure deduced or recognized with this method must be necessary supported by other considerations, taking into account the historical and social skill of the network, the cognitive organization of work, and different interdisciplinary aspects.

3.1.2 Justifying Necessity of SME Network Governance

The common suggestion for the above group of SME network models could be summarized as follows.

In any industrial frame, a *demand and supply network* is a "temporary network" of several SMEs, which decide to cooperate together in a common value-chain for a limited time horizon. In practice, it consists of a set of different enterprises, able to produce different parts which could be utilized in a common family of final products, and to apply complementary production programs, planned together, for a common industrial goal. This SME organization is usually characterized by "cooperation agreements" filled out by enterprises interested in cooperating together but only for a finite time period, and in such a way to involve only a part of their own core business. So, the resulting new supply chain can have a finite lifetime and it does not completely reduce the autonomy of any component firm, because each one can still produce items for proper clients, and then operate in a proper market segment. More precisely, all the enterprises which agree to be included into a demand and supply network and then be active inside the same supply chain, must sign an agreement to cooperate together in defining common production plans for specific products; obviously, they could also maintain their independence and autonomy for any other production.

In formal terms, a demand and supply network can be viewed as a *chain of production stages*, each stage containing either a SME or a set of parallel SMEs. Each SME, in turn, is managed by a proper autonomous decision-maker (theoretically denoted as an *agent*), who aims to cooperate but also wants to obtain the best profits for his own enterprise. Each production stage is connected to the upstream stages and to the downstream ones through a market place: it means that each SME can negotiate contracts for producing goods with downstream (buyer) SMEs as well as contracts for acquiring materials with upstream (supplier) SMEs. This *negotiation opportunity* is a qualifying character of any demand and supply network.

Since each SME aims to gain its own best income, it utilizes the demand and supply network to which it belongs, as a frame within which a "good negotiation" can be performed. "Good negotiation" means that an agreement between each pair of "*consecutive agents*" (i.e. agents belonging to two consecutive stages of the chain) can be found such as to satisfy both, because both aim to be cooperative, but at the same time, want to make profits: then the desired agreement should assure a sufficient income for both of them.

In practice, considering the complete working sequence of a final product to be supplied by the SME network, each SME plays the role of "buyer", which purchases materials in order to apply its own manufacturing operations and transform materials into product components, to be sold to the downstream market place. Therefore, the functions of any component SME are: purchasing, manufacturing and then sale of the produced items.

As an example, assume that the demand and supply network is composed of four SMEs, belonging to two different stages corresponding to two working

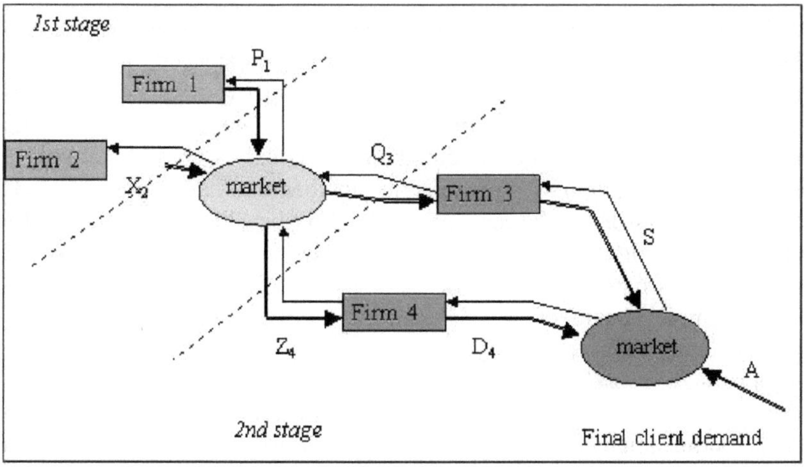

Fig. 3.5 A multi-stage demand and supply network

phases, and by a final client. Two market places, one located between the first stage and the second one, and the other at the second stage output, are realistic representations of the interactions of SMEs together.

Referring to Fig. 3.5, at the material input stage two enterprises (denoted by the indices 1 and 2) produce parts to be sold to the two SMEs belonging at the downstream stage (indexed by 3 and 4). These last ones apply final operations to transform items into final products, to be sold to final clients, modelled in terms of a known exogenous demand (denoted A in Fig. 3.5).

Accounting for the management issues in a realistic way, the governance of each SME has to be modelled by a production optimization strategy where maximum possible values for production volumes and price are searched for. This individual governance model corresponds to the goal of each SME to maximize its own profit, that means the difference between sell return and production cost, by using the highest possible production rate and leaving as small as possible their buffers.

But, if each SME governance adopts this same type of production management strategy, a full competition occurs among all SMEs: this situation forces the network to die.

This conclusion could be dramatic: if no control of the SMEs competition is applied, no profitable interactions among the SMEs themselves can exist.

Let us consider some additional constraints on the interactions among SMEs which could help to control competition.

A first constraint should be referred to the production flows: all items produced by SMEs of the first stage should be purchased by the second stage ones, and the demand for final products received from the client should be fully satisfied. This is a stationary situation in which, considering each one of the two stages, the production volumes of the SMEs located there could be balanced according to their respective capacities. If so, each SME will receive an income proportional to its own efficiency.

A second constraint can be related to the prices of items negotiated in the two market places: price and cost, at each market, should be balanced. This second constraint shows how the networks should operate in order to regulate the financial negotiations between concurrent producers: prices requested by the SMEs belonging at the same stage should be maintained as equal as possible otherwise a "dumping" situation could arise, thus causing lack of trust (and making the network weaker and weaker). Indeed, any price variation between the SMEs in the same stage reflects on price imbalances in the interactions with the other stages, both upstream and downstream.

These quite simple considerations, which can also be translated into mathematical formulations (Villa et al. 2005), can be viewed as practical proof of the necessity of a *network coordinator*, with at least two main management tasks.

On one hand, the network coordinator should monitor and control any production volumes' perturbations. In case the volume produced by a stage increases (e.g. owing to extra profit desire), the stage should be charged for an additional coordination cost proportional to the extra volume offered by suppliers to the downstream buyers (in the form of a "vertical re-balancing volume cost").

Since the extra volume could be due to a decision by one of the SMEs in a stage, a similar balancing action has to be assured within the same stage: the SME causing the volume imbalance must be charged for a coordination cost depending on the excess of production (now, in the form of an "horizontal re-balancing volume cost").

On the other hand, the network coordinator should also monitor and control any price perturbation. In case of an increase of the price of an item offered by a SME, a horizontal re-balancing price cost should be applied, in the same form as sketched above for the extra volumes.

It can be noted that price coordinating costs are strictly interrelated. Any price variation between two SMEs of the same stage, indeed, immediately reflects on prices imbalances in the other stages, thus making necessary the perturbation smoothing action of the coordinator.

3.1.3 *Benefits from Cooperative Management in Enterprise Networks*

Among the benefits for an enterprise belonging to a cluster, is the stability of a cluster, also pointed out by other authors, who considered the regional cluster from an evolutionary point of view (Becattini 1990). The reason resides in the collaborative activities performed inside a district which cover different aspects of the network existence and which not always have outcomes directly measurable in terms of economic profits. It is apparent that the study of only one aspect of the cluster activities can lead to hasty conclusions.

There is a large amount of literature on successful case studies in which collaboration gave a boost to both the quality and the profits. It is important to note that collaboration is seldom reduced to logistic but is applied to several enterprise processes: design, marketing, distribution, personnel training.

Examples of successful collaboration are: the widespread practice of developing new parts by co-design between the client and the supplier; the marketing practice of advertising a brand common to all the firms in a regional cluster, sometimes by obtaining that the product be manufactured exclusively in that region (Appellation d'Origine Controlé in France, Dominio d'Origine Controllata in Italy); the creation of scientific or technologic parks composed by a number of enterprises located in a territory having strong links with a local university or research centre.

A thorough definition of collaboration was given by the studies of Huxham (1996). Collaboration literally means "working together". The collaboration among the different companies belonging to a network, in order to optimize the management of the supply chains, leads to recognized strategic assets (Bergman and Feser 1999). The successful examples of collaboration reported in literature regard the collaboration in the implementation of business processes related to the supply chain management: planning, sourcing, production and delivery (as defined by the Supply Chain Council).

In the framework of the Collaborative Planning, Forecasting and Replenishment (CPFR) Committee a White Paper was written (Nix et al. 2004), which identifies the requirements for a successful collaboration among independent firms:

- Jointly managed business processes
- Standards for the sharing of information (data formats)
- Methods of integrating the results of this collaboration into the operational systems of both distributors and suppliers
- Key performance measures for joint supply chain activities

In order to assess the importance and the effectiveness of collaborative mechanisms it is also important to observe that industry networks implement different degrees of collaboration. It is possible to classify the following collaboration levels:

- *Ad hoc.* Collaboration does not go beyond the traditional customer supplier relationship.
- *Defined and linked.* Collaboration focuses on operational issues and limited to collaborative planning, forecasting and replenishment of materials and capacities, i.e. supply chain management.
- *Integrated and extended.* Collaboration at a strategic level where integrated and coordinated strategies lead to strategic synergy, i.e. extended and virtual enterprises.
- To this spectrum of maturity, we would also add clusters, which represent integrated collaborations that also include supporting infrastructures.

Therefore "collaboration" is a term that includes a wide spectrum of different situations when applied to the industrial world of enterprise networks.

The importance of information sharing for collaboration and cooperation is stressed in some studies (Gavirneni et al. 1999). These studies developed and exercised a minimal cost model to compare three levels of information sharing:

1. De-centralized control: there is no information sharing.
2. Coordinated control: two adjacent nodes share their information about customer demand.
3. Centralized control: total information sharing exists.

The results of the study show a steady improvement in value for all members of the industry network as the level of information sharing increases.

A difficulty in the assessment process of the collaboration paradigm is the exact determination of the benefits obtained by collaboration inside industry networks. The matter should not be reduced to an economic issue, but it is difficult to appreciate quantitatively the number of improvements obtained through collaboration. A reasonable solution can be the value analysis of networks collaboration by finding reliable value propositions, which refer to a customer-oriented approach (Bititci et al. 2004; Lewis 1990). The study of Bititci analysed the value creation in collaborative networks by extending the concept of value propositions from the single enterprise to an upper level, including supply chains, extended enterprises, virtual enterprises and clusters.

As a matter of fact if collaboration means exchange of information one should gather data about the flow of information: the more intense the flow, the more collaborative the network. In case (unrealistic) it would be possible to gather all the information exchanged in a network, it should be apparent that a mere quantitative analysis is insufficient. The quality and the reliability of the information exchanges is far more important than quantity.

3.2 Addressing Collaboration in Industrial Networks

B. Caroleo – Politecnico di Torino

According to the features exposed in the previous section, collaboration in industrial networks results as a prominent issue to use in analysing local productive systems. For this reason, the focus in this section will be on collaboration among enterprises in an industrial network.

3.2.1 Collaboration

Several definitions of collaboration exist in the literature (Appley and Winder 1977; Phillips et al. 2000; Wood and Gray 1991). As a matter of fact, the collaborative network (CN) phenomenon is described and interpreted in many different ways, depending on the background of the researcher.

Here, "collaboration" stands for a mode of "working in association with others for some form of mutual benefit" (Huxham 1996).

Specifically, organizational collaboration is defined as a "process in which organizations exchange information, alter activities, share resources and enhance each other's capacity for mutual benefit and a common purpose by sharing risks, responsibilities and rewards" (Himmelman 1992). This definition of collaboration is inclusive enough to include a wide range of collaborative activities (e.g. consortium, alliances, joint ventures, roundtables, associations) or also a firm's supplier relationships. Figure 3.6 shows an example of collaboration, referring to the joint venture established in 2006 between Panasonic and NEC. The synergy between NEC's communications and computing expertise and Panasonic's consumer electronics and audio-visual product experience, results in the development of a common software platform and to the development of new mobile handsets.

Sometimes collaboration emerges from a series of informal and unplanned relationships (Hakansson 1990; Von Hippel 1988). As a matter of fact, "many aspects of business relationships can never be formalized or based on legal criteria" (Gadde et al. 2003).

Collaboration can be studied from several perspectives, highlighting different features. Researchers working on learning and innovation agree in stating that collaboration can facilitate the creation of new knowledge and the transfer of existing one. For example, Anand and Khanna (2000) find evidence of large learning

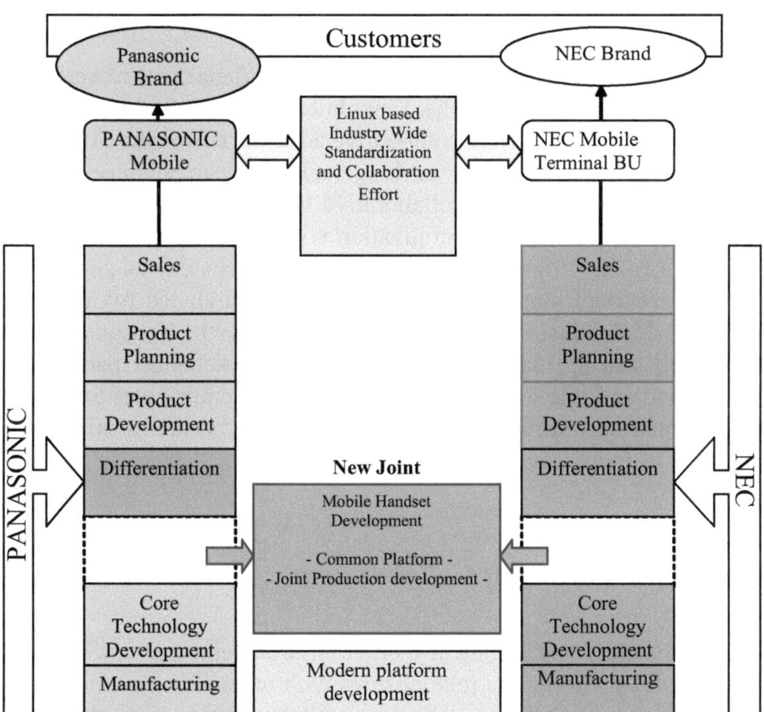

Fig. 3.6 An example of collaboration: the joint venture between Panasonic and NEC (adapted from Panasonic)

effects in managing joint ventures. According to Kale et al. (2000), one of the main reasons for which firms participate in alliances is to learn know-how and capabilities from their alliance partners. Powell et al. (1996) argue that linkages and the resultant collaboration networks are key vehicles through which firms obtain access to external knowledge.

According to network perspective and social capital view, collaboration can influence the structure of inter-firm relationships (Wasserman and Faust 1994; Wasserman and Galaskiewicz 1994). For instance, Wasserman and Galaskiewicz (1994) discuss the effects of inter-firm network structures from different perspectives: organizational power, performance, strategic decision-making and non-economzic activities.

Works on political aspects of cooperation stress the importance of collaboration to make some enterprises more central, increasing the influence over other organizations (Burt 1992; Hardy and Phillips 1998; Knights et al. 1993). Using examples from a study of the UK refugee system, Hardy and Phillips (1998) argue that collaboration is a strategy used by organizations in order to try to manage the inter-organizational domain in which they operate.

Firms within the network are not free to act according to their own aims: "they do not operate in isolation from others, or in response to some generalized environment as 'one-against-all'. Instead, each companies' considerations and actions can only be fully understood within a structure of individually significant counterparts and relationships" (Hakansson and Ford 2002).

Information sharing and exchange is recognized as a crucial part of collaboration. In their analysis of *keiretsu* networks, Lincoln et al. (1992) find in communication exchanges a key to increase the competitive advantage of firms. Also Tomkins (2001) argues that "development of long-standing deep alliances implies sharing information, and working out collaborative futures". Sabel (1996), facing the issue of the decision-making de-centralization within and among firms, states that enterprises have to adopt disciplines requiring continuous exchange of information about organizational improvements. A survey research led by Myhr and Spekman (2005) on 157 industrial relationships shows that "by constant interaction and information sharing via electronically mediated exchange, partners experience a closer bond and this serves to reinforce trust that contributes to collaboration". Radhakrishnan and Srinidhi (2005) report that the "value of information exchange is derived from improved resource coordination".

3.2.2 Reasons for Addressing Collaboration

In an increasingly dynamic and turbulent market environment, a firm's ability to develop and successfully manage its relationships with other firms is emerging as a key competence and source of sustainable competitive advantage (Batt and Purchase 2004). For instance, Zeller (2002) investigates the importance of collaboration in facing the strategic and organizational changes which involved two major Swiss pharmaceutical companies.

The reasons for cooperating are clear: enterprises should collaborate to gain access to combinations of resources that allow firms to perform activities they could not do alone. For instance, Fig. 3.7 shows the result of a survey conducted by ATP (Advanced Technology Program) in the manufacturing area to understand the motivations and impacts of joint venture collaborations. The findings show that collaboration among participants brings benefits from the pooling of resources and knowledge of their partners.

Furthermore, enterprises may want to reduce their risks and uncertainties or to gain knowledge from other partners, in order to reduce information costs and improve profits (DeBresson and Amesse 1991).

Collaboration takes a variety of forms, including creation and sharing of knowledge about markets and technologies (via joint research activities), setting market standards and sharing facilities (such as distribution channels). Delapierre and Mytelka (1998), for instance, illustrate an example showing how technological partnerships between firms have become prominent. As a matter of fact, the number of technological collaborations in the area of biotechnology rose from an annual average of 63 in the 1975–1979 period to about 536 per year in the 1986–1989 period. This empirical work has also highlighted certain structural features of collaborative activity: typically, collaboration ties are bilateral and are embedded within a wider network of similar ties with other firms.

Gulati et al. (2000) state that the networks of relationships in which firms are embedded profoundly influence their conduct and performance. As Gadde et al. (2003) argue, "relationship building may be the most important resource development process in any company"; furthermore, since networks potentially provide the firm with access to information, resources, markets and technologies, relationship building is also the source of a sustainable competitive advantage (Batt and Purchase 2004).

Network theory conceptualizes organizations as embedded in networks of linkages, which both facilitate and constrain their actions and shape their interests

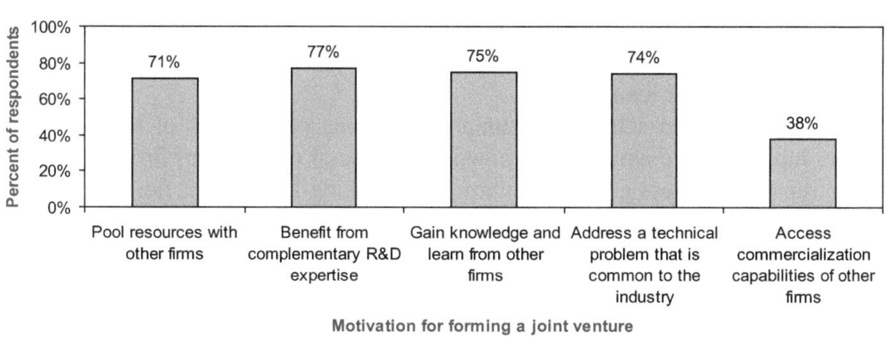

Fig. 3.7 Results of a survey on joint ventures (adapted from the Survey of ATP Joint Ventures, Advanced Technology Program 2006)

(Rowley et al. 2000). For example in his work centred on embeddedness, Granovetter (1985) argues that small firms in a market setting may persist because a dense network of relations connects them, reducing pressures for integration.

Proponents of a network perspective claim that the most significant aspect of an organization's environment is the set of other organizations with which to interact and the patterns of relationships among them. As Barley et al. (1992) argue: "not only are the organizations suspended in multiple, complex, and overlapping webs of relationships, the webs are likely to exhibit structural patterns that are invisible from the standpoint of a single organization caught in the tangle". These structural patterns, and the positions of organizations within them, have a significant impact on the degree to which organizations are able to control their own actions and influence those of others.

3.2.3 Examples of Collaborative Networks

Collaborative networks (CNs) are complex systems emerging in many forms in different application domains: the various manifestations of CNs have been studied by different branches of science, including computer science, computer engineering, management, economy, sociology, industrial engineering and law.

Pouly et al. (2005) study the most significant success factors of CNs, investigating two existing SME networks: Swiss Microtech (screw machining industry, producing parts for the automotive, medical, space and telecommunication branches) and USCO (suppliers to the machine industry, producing parts and services mainly to local machine manufacturers). The result is that in the case of competitor enterprises (same industrial branch), collaboration often consists in:

- Common purchase of raw material, machines, equipments or services
- Common marketing (e.g. participation in technical fairs) or sales activities
- Exchange of experiences about market trends and technical subjects

Asheim and Isaksen (2002) show how industries in two of the three Norwegian regions they studied have close inter-firm cooperation. In both regions, collaboration occurs both within the region and externally with technical research institutes in Norway or internationally.

Industrial-oriented scholars document numerous other types of CNs. For instance, Japan is one country well-known for its use of networks of firms, including the keiretsu, a business group of enterprises, working in different fields, linked by cross ownerships, a grid of relationships and other links not only official but also informal by belonging to a group. As an example, Fig. 3.8 shows the "Big Six" keiretsus and their network of relationships.

These networks form collaborations that are said to "reduce costs and risk, facilitate communication, ensure trust and reliability, and provide insulation from outside competition" (Lincoln et al. 1992). These networks are credited with hav-

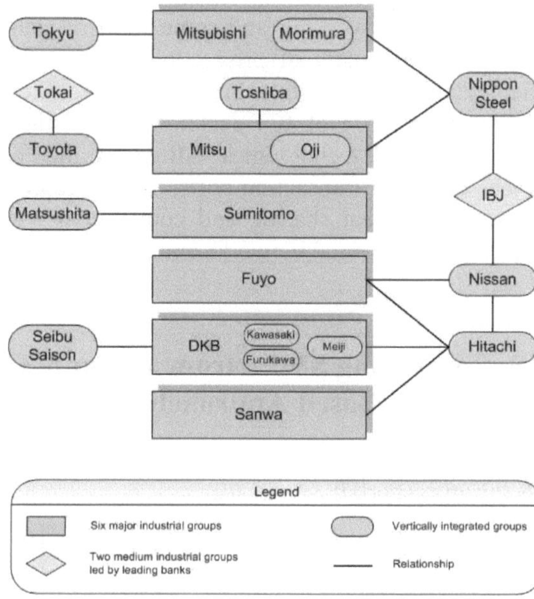

Fig. 3.8 Linkages of relationships among the Big Six keiretsus

ing allowed Japanese firms in the past to undertake risky, low-profit-margin, high-growth ventures (Fig. 3.9).

The need for collaboration had been recognized also in India; there is an active effort to link the various manufacturers with each other, through formal means. For instance, a field study of the SME jeans cluster in Bellary, Karnataka (Biswas et al. 2007) has shown that there is a lot of informal networking and interaction among the merchant manufacturers. Various activities have to be outsourced and

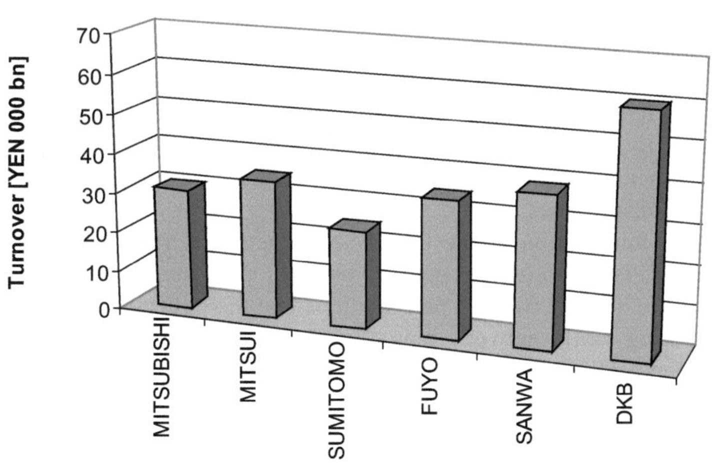

Fig. 3.9 Turnover of the Big Six keiretsus (adapted from Toyokeizai Data Bank, Japan Fair Trade Commission 1993)

this requires networking and some form of sharing. Even sharing of orders and sub-contracting parts of production, when individual units are unable to meet client demands, is possible through meaningful collaboration among clusters.

The importance of ICT technologies is highlighted by Biswas et al. (2007): adoption of ICT technologies will enable the enterprises in a cluster to integrate processes, lower transaction costs, and hence greater collaboration. Moreover, the authors assert that sharing and cooperation within the studied network can be improved with formal networks among the players.

3.3 Improving SMEs from a Supply Chain Perspective: a QFD-Based Approach

M. Barad – Tel Aviv University

Among one of the interests of the CODESNET project was the development of analysis methods useful for both supply chain and SME networks. To this aim, the quality function deployment (QFD), a product design quality technique, can be used for extracting and prioritizing the needs for improvement of a small sample of SMEs and for translating them into topics/tools of SCM evaluation. This section is devoted to illustrate how the QFD approach could be a significant tool for supply chain management (SCM) development.

3.3.1 The SME Situation in Israel Motivating the QFD-Based Approach

In contrast to the situation in many European countries such as Italy, Germany, France, Poland, Greece and Hungary, where the industrial evolution in recent times has been characterized by an agglomeration of groups of SMEs, in Israel it has not been possible to identify regional or other forms of organized SMEs *clusters* and their best practices. It seems that typically, SMEs in Israel do not have common organizational structures.

Hence, the Israeli contribution to the CODESNET activities related to SME networks collection offers a different perspective on this topic. The focus is on the *interaction* between the SME network representations and the collection of scientific reports concerning SME network main features and development, from a methodological perspective. The material collected and analysed during the CODESNET project led to the idea of describing the application of the QFD methodology for prioritizing supply chain-oriented concerns/improvement needs of SMEs and for linking them to relevant topics of SCM scientific knowledge.

The main objective of the section here is to analyse the improvement needs of SMEs from a supply chain perspective and detect the most relevant topics/tools of SCM scientific knowledge for responding to these needs.

The questions to answer are the following:

- How to *extract* and *prioritize* network-oriented improvement needs of SMEs?
- How to *classify* topics of SCM scientific knowledge?
- How to *prioritize* the classified topics of SCM scientific knowledge most relevant for improving the SMEs network performances?
- Do SME managers/engineers have the *knowledge and capability* to select the most relevant SCM scientific tools for responding to their supply chain (SC) improvement needs?

To answer the first question, empirical information has been gathered on supply chain improvement needs based on concerns expressed by interviewed managers/ engineers in a small sample of SMEs in Israel. The second question has been addressed through a structured compression of the vast literature on scientific contributions to SCM research and practice. To provide a methodological frame for linking the appropriate topics of scientific knowledge to the companies' improvement needs (the third question) the QFD, originally a product design structured technique for translating the customer's desires into product technical characteristics, has been adapted to the above specific purpose. The fourth question has been addressed by comparing the prioritized topics of scientific knowledge (QFD output) with the prioritized topics of scientific knowledge based on the interviewees' opinions (see Sect. 3.3.3).

The section is organized as follows. Section 3.3.2 presents scientific contributions to SCM in a nut shell. It is followed by a description of the methodology which includes the QFD technique, the pilot survey and the questionnaire. Then the findings are presented in terms of improvement needs and prioritized topics of SCM scientific knowledge. The section ends with a discussion of the results and suggestions for further empirical research.

3.3.2 Scientific Contributions to SCM Knowledge in a Nut Shell

The *evolutionary path* of SCM can be traced over the past fifty years. Many disciplines such as logistics, marketing, operations research, operations management and economics have contributed to its development. *Logistics* pioneered the concept of integrated logistics which became SCM. *Marketing* contributed to the idea of low pricing and postponement (late product customization), each being respectively linked to a different strategic competitive advantage of the SC: low cost versus quick response. *Operations research* and *operations management* focused on multi-echelon inventory models including vendor managed inventories (VMI), plant and distribution location, order allocation schemes, enterprise resource planning and distribution resource planning.

The recent increased motivation to research SCM, leading to an incredible proliferation of studies and publications was due to two types of factors: coercing factors and enabling factors. Among the key factors that according to literature *coerced* the development of supply chain management were: new customer service requirements, global competition, high logistics costs and the need to improve financial and logistics performances. Lately, attentiveness to the environment became an additional coercing factor. Highly developed communication and information technology can be considered major *enabling* factors of the SCM methods.

We deem that the key issue in SCM is dealing with the complexity of this topic as related to (1) studying and classifying existing supply networks and (2) studying and classifying existing scientific knowledge on how to manage these networks and analyse their performances as outcomes of decision making and method application. From the CODESNET perspective, as shown in Chap. 4, these two major different directions of study are respectively addressed by the Virtual Laboratory (i.e. the catalogue of SME networks' descriptions and industrial data formatting) and the Virtual Library (i.e. the archive of scientific reports concerning main issues and design/management problems of SME networks).

3.3.2.1 Classifications of Existing Supply Networks by Main Networking Activities

Several researchers attempted to build taxonomies of supply chains/networks by classifying them according to their main networking-oriented activities (motivation, partner selection, resource integration, information processing, knowledge capture, decision making, coordination, risk and benefit sharing). Harland et al. (2001) distinguished four types of supply networks resulting from combinations of two major factors: the degree of supply network dynamics and degree of focal company supply network influence (dominant partner). It seems worthwhile to mention that supply chain networks related to the sampled Israeli SMEs here, exhibited a high degree of focal firm supply network influence (at least one dominant partner big company in each SC).

3.3.2.2 Supply Chain Management Methods

Examining the vast and continuously growing literature on these methods, it has been decided to logically decompose this complex subject into key topics at two hierarchical levels and to base a critical analysis/discussion on a limited exploration domain, i.e. a few well-known references.

In their book *Designing and Managing Supply Chains*, Simchi-Levi et al. (2000) refer to three main topics: design, control and operations of SCM. According to these authors the objectives of SCM are reducing costs and improving service levels. By interpreting improvement of service levels as quick response, these two objectives represent the two *different* competitive advantages of SCM: low costs and quick response. SCM key issues detailed in the eleven chapters of the book are: "Logistics Network Configuration", a design decision, followed by "Inventory Management" and the "Value of Information" related to planning and control decisions. Much later in the book appears "Customer Value in the Supply Chain", a topic which is definitely related to supply chain design and strategy. It can be observed that in this book the term "coordination" is solely related to coordinated product and supply chain design.

Chopra and Meindel (2004) structured their book to consist of three decision phases: strategic framework and design (long term: parts 1 and 2), planning (medium term: parts 3 and 4) and operations (short term: parts 5 and 6). The objective of SCM becomes maximizing the overall value generated by the supply chain/network. The strategic framework highlights designing the supply net for achieving strategic fit, i.e. the capability to support the net ability to satisfy the targeted customers and their preferences: low cost or quick response. Both parts 5 and 6 address relationships with partners. Part 5 focuses on sourcing, transportation and pricing, while part 6 focuses on coordination and technology (information and e-business) emphasizing the negative effect of the lack of coordination on performance.

Other major references (Tayur et al. 1999; Shapiro 2001) stress the role of information technology (IT) in SCM and the need to differentiate between a bottom-up approach to IT (transactional IT) expressed by enterprise resource planning (ERP), material resource planning (MRP) and distributed resource planning (DRP), and a top-down approach to IT (analytical IT) expressed by strategic and tactical production planning, logistics operations, production and distribution scheduling.

De Kok and Graves 2003 seem to realize the undeniable importance of coordination and present it as a topic at the highest hierarchical level. They structure SCM in terms of design, coordination and operations. Design is associated with strategy while coordination is associated with tactics. Under the design topic, flexibility (manufacturing and product design-oriented) is strongly emphasized. Coordination comprises of contracts (which in Simchi et al. is considered a strategic topic), information sharing and tactical planning. Inventory is presented among the SCM operations but also appears as a design topic in terms of "safety stock and SC configuration", being related to the distribution strategy (warehouses location) of the SC.

3.3.2.3 Schematic Classification of Supply Chain Management Scientific Knowledge

Following the above discussion our suggested taxonomy blends several structures. As detailed below, we emphasize the overall role of *coordination* as a key topic and the importance of *flexibility* at a strategic level (both in design and in manufacturing), as suggested by De Kok and Graves. At a strategic level we also stress the significance of *strategic fit* as in Chopra and Meindl. At a tactical planning level we associate inventory management with information sharing as in Simchi-Levi. In our classification coordination addresses relationships among partners and the technology tools for achieving it. Accordingly, the highest hierarchical level is comprised of three key topics:

1. Strategy/design
   - Supply chain structure/design
   - Strategic fit (design to match the selected competitive advantage of the SC)
   - Flexibility in design and manufacturing
   - Distribution strategy (centralized versus non-centralized)
2. Tactical planning
   - Tactical/aggregate planning models (planning demand and supply)
   - Planning inventory (including information sharing: analytical IT)
   - Models for cost/profit distribution
3. Coordination
   - Outsourcing
   - Collaboration and contracts
   - Transportation systems
   - Information technology: transactional IT such as ERP
   - Problem solving assistants
   - Integration through e-commerce

3.3.3 Methodology

3.3.3.1 Quality Function Deployment (QFD)

QFD is a product design methodology whose essence is to extract the customer needs or desires, expressed in his/her own words (customer's voice), to translate them into technical product quality characteristics and subsequently into components' characteristics and operating decisions (see, e.g. Akao 1990). The voice of the customer represents a set of customer needs where each need has assigned to it a priority, which indicates its importance to the customer. Griffin and Hauser (1993) consider the gathering of customer information, extracting their needs, to be a difficult qualitative task, typically carried out through interviews and focus groups.

The QFD technique was developed in 1972 at Mitsubishi and, since the 1980s and the 1990s, has been gradually and successfully adopted by US and Japanese firms (see, e.g. Bossert 1991 and King 1995). Typically, the approach is described in terms of a four-phase model, consisting of four successive stages or matrices. Customer preferences (priorities) with respect to product attributes (the WHATs), are *input* to the first and best known QFD matrix, known as the House of Quality

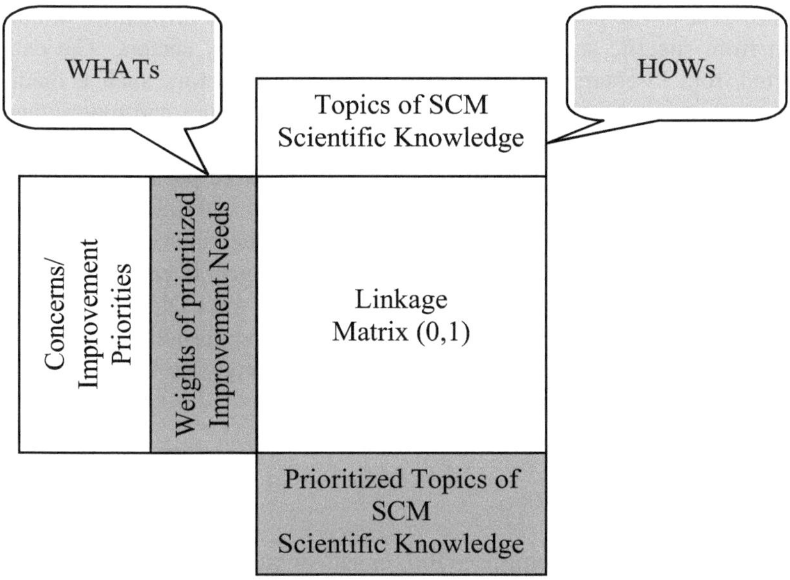

Fig. 3.10 Simplified QFD matrix for linking SCM knowledge to SMEs improvement needs

(HOQ). The HOQ *output* is the prioritized product technical characteristics as determined by the product designers for responding to customer desires (the HOWs).

To extract the supply chains improvement needs here the method developed by Barad and Gien (2001) is applied. It is assumed that the SC improvement needs stem from *concerns* that express unsatisfied needs (see Fig. 3.10). Customers are the interviewed managers/engineers in the sampled SMEs, the WHATs are their supply chain-oriented concerns with their respective gravity, and the HOWs are the topics of scientific knowledge, which in our opinion are most relevant for responding to these concerns.

To prioritize the improvement needs of a supply chain the importance attributed by the enterprise to each main process and the gravity of its associated concerns has been considered (see the Questionnaire below). The higher the importance assigned to a process and the higher the gravity of its concerns, the higher are its improvement needs. This concept is equivalent to the "importance-performance matrix as a determinant of improvement priority" developed by Slack (1994).

3.3.3.2 The Pilot Survey and the Questionnaire

During the first year of the CODESNET project a survey of nine companies from two sectors was carried out: electronics and food industries. It was comprised of five companies from the electronics sector and four companies from the food sector. Among them, there were five SMEs (less than 150 employees and $50 million annual sales) and four larger enterprises.

In the third year of the project the survey continued but had difficulties in collecting data from specific sectors for analysing the results by sectors. The data were collected from 13 enterprises representing a variety of sectors, such as medical products, cosmetics, metal products, shoe industry, electronics and food industry. Among them, there were eight SMEs and five larger enterprises.

The investigated firms in the two stages of the survey were positioned at various points in the value process, from product design to raw materials and distribution. Data were collected through semi-structured face-to-face interviews with SC managers and/or material managers and/or operations managers in each enterprise. Basically, the same questionnaire as the one designed for the CODESNET website, was used in the two surveys but, to be better understood, formulation of some questions was simplified in the second survey. The findings are based on data collected during the two stages.

3.3.3.3 Interview Questionnaire

The interview questionnaire is comprised of four sets of questions:

1. The first set was of a general nature, supplying *background data* on the investigated enterprises such as sector, main product, total number of employees and annual sales. To get a quick look at the companies' economic status we asked an additional question: "What is the bottle neck for increasing sales?" Specifically the interviewees had to select one among the three alternatives: demand, manufacturing capacity or suppliers.
2. The second set provided information on their *position,* i.e. function(s) performed by the enterprise along the value chain of their main product, e.g. new product development/design, marketing, manufacturing/assembly, raw material suppliers, distribution/retail.
3. The questions in the third set were used to determine the *supply chain improvement needs* by providing two information items:
   - The *importance* level (from 1 (not important) to 5 (most important)) associated with the improvement of each of the six main processes: product design, manufacturing, supply (procurement/delivery), design of the supply net, planning the demand supply of the supply chain, collaboration/integration.
   - Concerns/problems occurring in each of the above main processes and their gravity level (from 1 (minor) to 5 (critical)).
4. The fourth set of questions asked the interviewees to prioritize fourteen topics of *scientific knowledge on SCM* as judged by their potential to improve the performances of the six basic processes mentioned above. Again, the priority level was from 1 (not significant) to 5 (very significant). The list of topics represented the schematic classification detailed in the previous section.

This information was used to compare the priority levels attributed to the scientific topic by the interviewees with the priority levels obtained through our application of the QFD method, i.e. translation of the prioritized concerns into prioritized topics of SCM scientific knowledge.

3.3.4 Findings

3.3.4.1 Improvement Needs

1. Bottleneck for increasing sales

In 68% of the 22 enterprises, *demand* was the bottleneck for increasing sales, in 23% it was *manufacturing capacity*, while the in the remaining 5% of the enterprises it was related to *suppliers*. Enterprises that complained of shortages in manufacturing capacity were from the electronics, cosmetics and medical equipment sectors. Eventually, by improving their manufacturing efficiency they could make better use of their existing capacity.

2. Processes to be improved

Processes whose improvement was considered highly important by more than 50% of all companies were: the SC collaboration process, supply and manufacturing. Improvement of the product design process was considered highly important by about 40% of all enterprises. Less than 20% of all enterprises considered that improving SC design and/or SC planning was highly important.

3. High-gravity concerns and prioritized SC improvement needs

Table 3.1 presents a partial view of some SC high-gravity concerns (4–5 gravity level) matched to topics of SCM scientific knowledge from Sect. 2.3, which in our opinion provide responses to these concerns.

A list of prioritized SC improvement needs was obtained by considering all high-gravity concerns (4–5 gravity level) related to important processes (4–5 importance level). The total scores (weight) of prioritized SC improvement needs appearing several times on the list (several enterprises formulated it) were calculated by summing up their gravity scores.

3.3.4.2 Prioritized Topics of SCM Scientific Knowledge

The prioritized SC improvement needs were translated into prioritized topics of SCM scientific knowledge (from the topics' list in Sect. 3.3.2) by applying the simplified QFD methodology illustrated in Fig. 3.10. To allow a simple comprehension of the results, the original QFD matrix representing *strength of relationships* between the needs and the responses to these needs has been simplified, by using binary (0,1) values. More specifically, it has assigned the value "1" to a topic that in our opinion could respond to any prioritized SC improvement need.

Table 3.1 Some SC high-gravity concerns matched to topics of SCM scientific knowledge

Strategic fit	Flexibility in design and manufacturing	Distribution strategy centralized/ non-centralized	Planning and managing inventory with information sharing	Outsourcing	Contracts	Transportation	Transactional IT	E-commerce
Rapid transportation is costly Quick reaction is needed but supply contracts are rigid	Small Batches Long set ups Product variety Changes in design Long time to market Modularity is needed Technological changes take time Short-term capacity shortages	Storage space problems related to contracts with customers Storage space problems related to contracts with suppliers	Not reliable information High inventory related to product variety Frequent changes in demand Inaccurate forecasting Seasonal demand Inventory managed by customers – not optimal High inventory of raw materials	Scarcity of resources creates dependability Lack of collaboration on product design Raw material late Unreliable supply date Lengthy delivery	High inventory due to rigid contracts Different agreements with different customers Lack of coordination on supply dates Storage space problems related to contracts	Rapid transport is costly No collaboration on transport along the SC	Old ERP not compatible with SC Human errors in input data to ERP	High inventory related to product variety Long time to market

Table 3.2 Prioritized topics of SCM scientific knowledge from two information sources

SCM scientific topic	Research-based QFD ranking	Direct evaluation customer ranking	Consensus	Discrepancy
Supply chain structure/design				
Strategic fit	7<	4		
Flexibility in design and manufacturing	3	7		Underestimated by customer
Distribution strategy				
Aggregate planning models				
Planning inventory and information sharing	2	1	√	
Models for cost/profit distribution				
Outsourcing	5	3	√	
Collaboration and contracts	1	6		Underestimated by customer
Transportation systems				
Information technology: transactional IT	4	2	√	
Problem solving assistants	7	5	√	
Integration through e-commerce				

The topic numerical evaluation (its weight) was the total gravity score of all the prioritized improvement needs (selected concerns) it could respond to. If several topics could respond to a particular concern, its score was taken into consideration in the weight calculation of each topic.

Table 3.2 presents the prioritized topics of SCM scientific knowledge ranked in descending order by two methods: (1) QFD translation of prioritized SC improvement needs (research-based) and (2) direct evaluation by interviewees (customer-based).

3.3.5 Discussion of Results and Further Empirical Research

The pilot study, which applied the QFD methodology in a simplified but innovative manner, reveals the most relevant topics of SCM scientific knowledge for improving the supply chain performances of SMEs. The five most relevant topics were: collaboration, planning and managing inventory with information sharing and risk pooling, flexibility in design and manufacturing, information technology and outsourcing methods.

By comparing these results with the customers' direct evaluation of these topics it can be concluded that SME managers have the knowledge and capability to rightly evaluate the advantages of using most of these tools. This regards the major effects of inventory management and the value of information, the enabling task of information technology and of the problem solving software assistants.

Two important discrepancies have also been detected in their evaluation: the role of *flexibility in design and manufacturing*; as well as the role of *collaboration/coordination and effective flexible contracts*, which were underestimated by SME managers.

In addition, the data collected during the first stage of the study showed differences between the electronics sector and the food sector. There has been limited empirical research both on classification of existing SMEs networking activities as well as on applications of SCM scientific knowledge to improve their performances. Further empirical research in these directions may reveal beneficial results.

3.4 SMEs in Supply Chain Networks: Cooperation and Competition

J. C. Hennet – CNRS-LSIS

The concept of a supply chain concentrates some major features of business organization in today's society. Typically, it represents a network of autonomous production units connected through a communication network for flow of information and a logistic network for flow of products. The worldwide evolution toward more specialization of activities and better communication and information systems, offers supply chains higher capabilities than those of centralized firms in terms of flexibility, responsiveness and costs. Aside from this evolution, the existing industrial configuration in the majority of the European countries is mainly made up of SMEs. Coexistence of these two realities results in raising the questions of their compatibility and future evolutions: Is it valuable for a SME to upset its organization and to give up its traditional clientele to become a minor actor on a market certainly more global but maybe more risky? And, if the answer is yes, how can a SME achieve these changes, and what support can it expect from the sharing of resources and the development of innovation programs in existing or emerging structures like districts and clusters?

3.4.1 The SCM Perspective

In the considered context of Collaborative Demand and Supply Networks (CODESNET), supply chains organize sets of manufacturing activities structured in stages, from supply in raw materials and components to delivery to customers. Different firms are involved in a supply chain and their production and procurement policies have to be coordinated to achieve value chain efficiency. To better focus on customers' requirements and information flows, a supply chain is generally analysed upward, starting from customers and end-product retailers, then looking at manufacturers, tier 1 suppliers, tier 2 suppliers, and so on. Typically, a supply chain network is not restricted to a single chain (in the graph theory terminology). Manufacturing flows may follow divergent and convergent paths, as in the cases of product diversification and assembly. Information flows may spread out much more over the network to create the largest set of possible production schemes, among which the best one could ideally be selected.

However, in practice, decisions on the structure of the supply chain cannot be considered optimal since they are taken locally by legally separated organizations and they are subject to time constraints of minimal duration of contracts and financial agreements.

Table 3.3 Performance attributes and level 1 metrics (adapted from the Supply Chain Council)

Level 1 metrics	Performance attributes				
	Customer facing	Internal facing			
	Reliability	Responsiveness	Flexibility	Cost	Assets
Perfect order fulfilment	×				
Order fulfilment cycle time		×			
Upside supply chain flexibility			×		
Upside supply chain adaptability			×		
Downside supply chain adaptability			×		
Supply chain management cost				×	
Cost of goods sold				×	
Cash-to-cash cycle time					×
Return on supply chain fixed assets					×
Return on working capital					×

3.4.1.1 Measures of Supply Chain Effectiveness

Several key performance indicators (KPI) have been proposed to measure the effectiveness of existing and prospective supply chains. In particular, the SCOR model (SCC 2000) proposes to match high-level metrics with supply chain performance attributes (see Table 3.3).

3.4.1.2 SCM Requirements

In the context of globalization, world-leading companies and international institutions such as the World Bank, recognize the importance of SMEs as economic actors and potential supply chain partners, mainly for their specific know-how and their strong potential in terms of employment. In OECD countries, SMEs account for 60–70% of employment and this percentage is generally much higher in less developed countries (GFP 2005).

In addition, the use of SMEs for outsourcing and sub-contracting has generalized to the point where the cost of purchased materials constitutes more than 50% of total sales for an average manufacturing firm (Subramaniam 1998).

As a counterpart of these assets, SMEs are often critically deficient in several respects. A list of 14 major obstacles to internationalization has been identified by UNECE (United Nations Economic Commission for Europe). They are listed below by decreasing order of importance:

1. Lack of entrepreneurial, managerial and marketing skills
2. Bureaucracy and red tape
3. Lack of accessibility to information and knowledge
4. Difficulties accessing financial resources/lack of capital
5. Lack of accessibility to investment (technology equipment and know-how)
6. Non-conformity of standardization, lack of quality awareness and lack of mutual recognition schemes
7. Product and service range and usage differences
8. Language barriers and cultural differences
9. Risks in selling abroad
10. Competition of indigenous SMEs in foreign markets
11. Inadequate behaviours of multi-national companies against domestic SMEs/ lack of government supply-supporting programs
12. Complexity of trade documentation including packaging and labelling
13. Lack of government incentives for internationalization of SMEs
14. Inadequate intellectual property protection

Some of these requirements, for example 3, 4, and 5, relate to the SME intrinsic weakness in a global market. However, this list also reveals the dominated status of SMEs in most supply chains. It is clear that requirements like 7, 8, and 9 probably emanate from multi-national groups, leading OEMs or institutions, in their attempts to remove supply-side constraints and connect their associated SMEs to targeted markets.

3.4.2 The SME Perspective

The difficulty for a SME to create the conditions for entering an important supply chain raises two main questions:

- What are the other alternatives for development and how to choose among them?
- If entering a large supply chain is the preferred alternative, how to progress toward meeting SCM requirements?

These issues will now be addressed from the three classical decisional levels of strategy, tactics and operations, with their associated time scales, as in classical hierarchical production planning systems (Bitran and Hax 1977). However, the distributed nature of supply chains modifies the type of models to be used at each stage. Also, decisions taken at upper decisional levels (respectively strategy, tactics) can only serve as frameworks and guidelines for lower decisional levels (respectively tactics, operations). At each level, several options are analysed through their advantages and drawbacks for SMEs.

3.4.2.1 Strategy for a SME

Diversification or specialization?
In the manufacturing sector, small business management must address the issues of raising capital, hiring competent qualified personnel, investing on industrial and administrative premises and building a reliable business plan. A key factor influencing these strategic decisions is the choice of products and markets, with two possible trends: diversification or specialization. Diversification offers the advantages of flexibility and adaptability of human resources and equipments, with the counterpart of higher costs and forecasts with shorter term. On the other hand, a well-chosen specialization may reduce costs by relying on dedicated equipment, specialized personnel and large series. Well-targeted products and low costs are determinant factors for taking part in global supply chains and penetrating large markets. However, the risks implied by a sharp specialization are very important: competition and changes in the market may cause the break of a contract, which generally leads to bankruptcy.

Various remedies can help specialized SMEs to overcome the risks of specialization. Better competence and a better capacity of adaptation of human resources can be achieved through continuing education programs and rotation of assignments to jobs and workstations.

Competition between suppliers of a dominant retailer
Retailers often face random variations both on the demand and the supply side. In such cases, Arda and Hennet (2006) show that it is generally more profitable for the retailer to procure from several suppliers rather than to use a single one. Similar results apply when the retailer is concerned with supplier default risk (Babich et al. 2007). An increase of wholesale prices for smaller delivered quantities could balance this trend, except that wholesale price increase is often very limited, in particular if the retailer is in a dominant position or if there is a strong competition between suppliers. A possible response in this situation is to develop a symmetric strategy, as described in the V-LIB paper from Sucky (2007) where a supplier dynamically selects his/her retailer. However, this would be an attempt to become a dominant partner, which is particularly difficult for a small enterprise facing larger ones.

Supportive business environment for SMEs
Point 4 in the UNECE list is often critical for a SME. It relates to lack of capital and investments. In some cases, the investments in expensive equipment can be shared between the SME and its partners in the supply chain, in particular its customer company or the SC dominant enterprise. Co-investment policies have not been much studied in the context of supply chains. A recent work on this issue is based on game theory and shows that the same equilibrium points can be obtained from Nash and Stackelberg approaches (Kogan and Tapiero 2007). This result seems to indicate the relevance of co-investment, whatever the relations of dominance within the supply chain.

Another possibility is mentioned in point 11 of the UNECE list. It is the use of government supporting programs. Several types of institutional supporting programs have been analysed on the CODESNET website.

SMEs should develop their openness and opportunism to penetrate larger markets by becoming members of supply chains. Such changes are difficult to achieve and should be anticipated. These observations have led to the emergence of associations of SMEs and to political incentives in several European countries. One of the major contributions of the CODESNET action is to provide examples of such associations and incentives.

In France, a new industrial policy was launched in 2004, to mobilize the key factors of competitiveness, and particularly the innovation ability, considered as the first. A pole of competitiveness relates to a particular industrial sector in a particular geographical area. It is an association of enterprises, research centres and education organisms engaged in a partnership incentive to generate synergies through innovating projects in the direction of a given market (see http://www.competitivite.gouv.fr). Financial supports are granted by national and regional political institutions to member companies developing innovative joint industrial projects. For example, the pole EMC2, on "Complex Metallic and Composite Structures" in the Loire Region, groups five world-class companies: Airbus, Chantiers de l'Atlantique, DCN (European leader in naval military construction), Bénéteau (world leader in yachts construction) and Auto Chassis International (equipment for the automotive industry). More than 1,000 SMEs belong to this pole. However, involvement in the pole is limited to innovating projects, often in a short-term perspective.

Industrial districts such as SNS (Supply Network Shannon) work in a totally different manner. They organize promotion of local SMEs through marketing events and exhibition in which larger companies are invited. In many cases, existing supply chains remain open to external collaboration, to allow for competition and maintain low costs. Then an SME may occasionally enter a supply chain as a venture partner and then build stronger links with the member enterprises.

A third alternative can be found, for example, in Italian districts where SMEs use their complementary skills to constitute the leading body of a supply chain. One example included in the CODESNET website is the "paper and cardboard district Lucca" which produces 40% of corrugated paper and 70% of tissue paper for the Italian market.

3.4.2.2 Tactical Issues

Tactical planning mainly deals with decisions on manufacturing and inventory policies, resource assignment and product flows with partner firms. SMEs experiencing integration in a supply chain must drastically change their planning tools, to meet the partners' requirements with limited increases in costs.

Decision making in supply chains

A typical tactical concern for supply chains is the organization of exchanges between adjacent firms. By convenience of terminology, the downstream firm will be called "the retailer" and the upstream firm "the supplier". Through their activity of transformation, assembly or sales, both firms contribute to the increase of the value chain. A key issue for this pair of firms is distribution of cost, profit and risk among them. It is directly related to decisions on whole sale and retail prices, dedicated capacity for production and storage, order and production lot sizes, costs of backorders and lost sales. Such decisions require agreements obtained through negotiation stages, in which partners elaborate contracts defining their commitments and terms of trade. It should be always assured that proposals, counterproposals and decisions should be based on scientific analyses rather than on intuitive grounds.

Many different models have been proposed to represent the local autonomy of production units within supply chains. As an organizational structure, a supply chain is a holonic system composed of different units having their own resources, methods and goals (Van Brussel et al. 1998). In this context, negotiation is necessary to achieve collaboration of partners with different and often conflicting objectives.

Such characteristics are of major concern in game theory, under the classical decomposition into cooperative (or coalitional) and non-cooperative games. For an introduction to game theory, the reader is referred to a textbook by Osborne and Rubinstein (1994). The models referred to as "non-cooperative" consist of players with different preference relations or utility functions. Their actions obey a strategy, which takes into account information (generally imperfect) on the other players' actions and preferences. Of particular importance in non-cooperative games theory are the problems of existence, uniqueness and characterization of an equilibrium set of actions, which is locally optimal for all the players. Under different rules, a game may lead to different equilibrium profiles: a Nash equilibrium if the game is symmetric, a Stackelberg equilibrium in the two-player case with a leader and a follower.

Models and contracts based on game theory
As noted in the work by Cachon and Netessine (2004), most supply chain models based on game theory use a non-cooperative approach. In this setting, a key result for de-centralized supply chains (supply chains with distributed decisions) is sub-optimality of Nash and Stackelberg equilibria with respect to a global economic criterion. As stated in the work by Cachon and Zipkin (1999), "competition reduces efficiency". A large number of studies have proposed contracts between suppliers and retailers to coordinate the system and reduce the optimality gap through cost and/or risk sharing. In Cachon and Zipkin (1999), a two-level supply chain is studied with a retailer subject to a stationary random demand and his supplier. Both firms have fixed supply cycle times and use a base stock policy in which an order is sent at each time period, to maintain the inventory position to a predetermined level S_i. Both partners have individual storage costs and share the retailer's shortage costs. The problem of each firm consists in determining the optimal stock level, S_i, called the base stock level, which minimizes his/her average cost. The authors show that this static game has a unique Nash equilibrium which differs from the optimal solution of the centralized system. Then, they propose coordination contracts characterized by linear transfer payments, which can drive the system to its optimal performance.

In a supply chain, a SME is often associated with a larger enterprise that has a dominant position. In this situation, the negotiation stage is likely to be represented as a sequential game with two stages, called a Stackelberg game. The first player, called "the leader", acts first by choosing a strategy among the set of possible ones. Then, the second player, called "the follower", is informed of this choice and selects one of the possible strategies. Such a game defines a relation of not fair strength in favour of the leader because by playing the first one, the leader can impose the equilibrium point which is ultimately reached. Along this line, Lariviere and Porteus (2001) characterize the Stackelberg equilibrium for a two-level supply chain where the supplier is the leader. He (she) decides of the wholesale price and the retailer (the follower), who faces a random demand, orders the quantity that maximizes her (his) expected profit.

The most frequent status of a SME in a supply chain is to be a supplier of a dominant retailer. A single product case is analysed in Arda and Hennet (2007). Each firm manages his local production/inventory control system while attempting to maximize his own steady state expected profit per time unit. The contract (p,b) defines a transfer payment from the retailer to the supplier. Parameter p is the wholesale price (paid by the retailer to the supplier), and the manufacturer charges a penalty $b1$ to the supplier per unit of component backorder and per time unit. Under the assumed based-stock policy, the main operational decision at each stage is the base-stock level S_i. It is shown that under the resulting contract, the two-stage supply chain reaches globally optimal running conditions with the maximal possible profit obtained by the retailer and the smallest acceptable profit obtained by the supplier.

What can be induced from these studies is that the terms of the contracts between a SME and the leading partners of a supply chain are essential to guarantee minimal revenue in this supply chain. The actual profit level will not generally exceed this minimal level. The basis of this statement is even stronger in the situation of competition between suppliers, described in the section on strategy.

3.4.2.3 Operations

The drastic changes implied by the embedding of a SME into a supply chain network should be based on improved operational practice. Placed in the position of a supplier within a supply chain, a SME should improve its efficiency and reactivity to conform to market requirements of response time, quantity and quality. For most SMEs, reductions in costs and improvements in quality require better methods and a better technology in finance, administration, design and manufacturing. Supply chains are characterized by a complex de-centralized decisional organization but a highly integrated operational system. As previously mentioned, decisions taken at the tactical level should be implemented, or at least used as guidelines, at the operational level. Such a strong integration has become possible within modern information and communication networks through the emergence of standards for the exchange of product model data (STEP, ISO-10303), progresses in business process management based on workflow (http://www.wfmc.org), and in process deployment methods using agents, business objects and business processes (http://www.omg.org). Communication and de-centralized implementation of decisions can be achieved through multi-agent systems (Shoham 1993).

As stressed in (Antonelli and Villa 2007), a major challenge for the effectiveness of the supply chain relies on the implementation by the component SMEs of KPIs consistent with the KPIs of the whole network. The sharing of the same standards in technology, information and communication is particularly critical in high-tech sectors, as illustrated by the example of the cluster "Aerospace and Environment" in the area of the Friuli Venezia Giulia region, in the northeast of Italy (see Sect. 2.5). In this cluster the leading company produces technological systems with components provided by SMEs. The number of ICT technologies adopted and used is thus very high and transversal to all the companies belonging to the cluster.

3.4.2.4 Some Final Remarks

The study of existing collaborative demand and supply networks in European countries has set into evidence the existence of relatively loose networks of SMEs organized at the regional level to share experience and resources and to improve their practice of modern information and communication tools. The origin of such districts or clusters of SMEs greatly differ among countries, with the two extreme situations of top-down incentives as in the French poles of competitiveness, and bottom-up grouping as in the districts of northern Italy. In any case, these networks play a key role in the evolution of traditional SMEs toward the emergence of global supply chains with higher specialization or specificity of local member firms. In the present situation, the organization in many SMEs is far from meeting the high standards of world-class companies. However, networks of SMEs contribute to bridging this gap and may also offer to SMEs alternatives other than joining a supply chain as a minor partner unable to correctly negotiate contracts with leading firms and subject to the high demand variations of competitive markets.

3.5 Trust in Demand and Supply Networks

T. Potinecke and T. Rogowski – Universität Stuttgart

The main suggestion coming from the previous descriptions of SME clusters and networks is that they contribute to strengthen the competitiveness of the participating enterprises, e.g. by increasing the innovation capability or by creating more efficient processes. In this context, *trust* has been identified as one of the most important key success factors of a cluster. This result, which in turn has been considered by other projects and authors (as it will be referenced in the following), came mainly from analysing the CODESNET catalogue of clusters for what concerns the question whether industrial partners use trust as coordination instrument in their network, mainly when some problems and difficulty which could affect the cluster survival occur.

Indeed, there also may be processes which lead to a weakening of the cluster in terms of reduced competitiveness, stagnation or economical decline of their performance. When clusters are characterized by a high specialization, this one leads to expertise, but it is always connected to vulnerability. In particular, technological discontinuity and fast technological changes or innovations may lead to situations in which specific cluster advantages are undermined. Economical changes and changing customer expectancies may contribute to problems in which specific advantages that had a positive effect on the cluster are outstripped since new forms of specialization have become necessary (Andersson et al. 2004). This development goes hand in hand with a certain inflexibility. Due to the loss of the ability to react fast and efficiently to changes in the environment, the probability of misinterpretations concerning future developments increases. Another important phenomenon concerning clusters is the appearance of “lock-in-effects”. These may occur if the cluster participants focus too much on local relationships and neglect external connections (Martin and Sunley 2003). Last but not least, it should be mentioned that further inefficiencies may appear when cooperating enterprises in clusters build up entrance barriers for new business foundations or settlements (Andersson et al. 2004).

In front of these types of potential weakness factors, the majority of the enterprises indicated that a successful partnership, clearly dependent on reciprocal knowledge and trust, can be the crucial condition to overcome any difficulty. Other enterprises indicated that the success of a partnership depend on detailed contracts. The mutual trust indeed, as well as honesty and openness between the partners represent the most important conditions and success factors of cooperation. Although in networks resources are asymmetrically distributed such as power, money, knowledge and experience and are used as arguments for the stabilization of the own position, partnerships presuppose a culture of trust (Pander and Wagner 2005).

All analysed enterprises hold the opinion that trust in partnerships positively affects the process of the cooperation relationship. They see particularly in the saving of time and expenditure the substantial advantages, which arise as a result of trust. Above average commitment and the transfer of sensitive information, which result in the case of the existence of trust, are also very highly rated.

Most of the analysed enterprises indicated that partners will operate without opportunistic behaviour, if they can identify themselves with the common goals of cooperation. On the other side there were some opinions that partners will behave trustworthily only if negative consequences and sanctions threaten them. The conception of a rationally structured cooperation in practice of organizations is unrealistic. Rather we can assume that trust represents a central variable for the structure and for the stability of cooperation and communication in organizations and that with a lack of trust the cooperation could fail. More substantially however than a contract is the common vision of the partners, which is to be developed.

The exchange of information and face-to-face relations were classified by many enterprises as the most important measures for the setting up of trust. Besides the clear distribution of authority and responsibilities, a trust culture was regarded also as very important, so that trust between enterprises can develop. Trust and reliability are indispensable, and in the measure in which trust-strengthening behaviours and cooperation atmosphere are experienced, usually the inclination to cooperation grows. Generally it applies that willing to cooperate is in the long run necessary, in order to get a successful cooperation (Ringle 2004). Structure of trust by regular communication leads in particularly to a genuine partnership. Contracts represent only a legal dimension. For the trust development the constant passing on of information is necessary. Reputation is seen as the most important criterion in order to judge the trustworthiness of potential partners, i.e. to judge potential cooperation partners regarding their ability and readiness for cooperation. Besides, external audit procedures also play a large role with the evaluation of the partners. This evaluation makes it possible for the partners to develop trust faster since already trustworthy behaviour of the partner was seen in the past. Past information represents the past behaviour of the partner. But past information can only give a limited guide to predict the future behaviour of the partners.

"Soft factors" like trust, honesty, openness and convergence of the cultures are substantial conditions of a successful partnership. This applies to enterprises with much as well as little experience with cooperation. Humans are regarded as central success factors in cooperation and relations management. These soft factors are even more important than organizational and technical aspects. For this reason the controlling of soft factors and therefore the relations management is becoming increasingly more important in order to achieve the hard targets of the network of enterprises.

The analysis developed during the CODESNET project revealed that partners in a SME network recognize the great importance of trust, however they are not always in the situation to use the trust. That is due to the fact that trust is not yet popular as a coordination mechanism. Enterprises must first get accustomed to the fact that soft factors represent the central success factors within a cooperation. How to cooperate with partners is the central question which one needs to know. One must be conscious that trust is more than an eliminating of distrust. Trust relations can be trained only mutually. The human factor will thereby in the future again play a substantial role for the success of enterprise-spreading cooperation. To recognize this factor and promote accordingly will be one of the central challenges for the responsible persons in the networks.

3.6 Training Networks in Ireland

D. Whelan and C. Heavey – Limerick

This section describes the development and current state of training networks in Ireland. Over the past 15 years Ireland has experienced unprecedented economic growth. Between 1990 and 1995 the economy grew at an annual average growth rate of 4.8% and, between 1995 and 2000, it averaged 9.5%. Growth rates have since been maintained in the 4–5% range, a level almost five times higher than the old EU15 average (IRELAND – Economic Profile August 2006). A major strategy of the Irish government in developing and supporting this growth is the continuous up-skilling of the workforce in which training networks have played a major part. Section 3.6.1 gives an overview of development of the current training strategy in Ireland evolved in the recent past. A major programme in Ireland in training is Skillnets and is described in Sect. 3.6.2. The following three sections describe three case study training networks.

3.6.1 Development of Training Strategy in Ireland

At the beginning of the 1990s the government commissioned a major review of industrial policy and performance paying particular attention to policy changes, which could be initiated or influenced by future government policy. The Industrial Policy Review Group reported its finding in 1992 in the Culliton Report (Culliton 1992). One of its recommendations was that there should be more on fostering industrial clusters around industrial segments and niches of national competitive advantage. It stated that this type of cluster approach was not supported by current development agencies policies.

Subsequently, the National Economic and Social Council (NESC; http://www.nesdo.ie/) commissioned a substantial study on industrial clusters in Ireland, examining in particular three Irish sectors: dairy processing, the music industry and the Irish indigenous software industry. Reports on these three case studies have been published by NESC (O'Connell et al. 1997; Clancy and Twomey 1997; O'Gorman et al. 1997). It was concluded that the three case study sectors studied where not leveraging to gain a competitive advantage that could be obtained from clustering (Porter 1990).

In 1995 the Science, Technology and Innovation Advisory Council (STIAC) Report (1995) concluded that the basic structural problem of Irish industry – the small size and scale of operation of most indigenous firms – must be directly tackled through a programme to bring enterprises together in cooperating groups. The approach was based on the Danish Network Co-operation Programme. In response a White Paper on Science, Technology and Innovation was issued and a six-month pilot programme on Inter Firm Co-operation Networks was initiated. The pilot program was managed by ForFás (http://www.forfas.ie/). During the six-month period, 17 new networks were formed. An evaluation of the programme found that the pilot programme had met all its objectives and it recommended the establishment of a long-term national program to enhance the networking capabilities of companies in Ireland. However this recommendation was never implemented.

Another development was the Plato initiative, which was developed in Belgium in 1988 with the objective of promoting regional economies. It is now a dynamic European-wide network, with over 8,000 SMEs and over 500 parent companies in Belgium, Holland, Germany, Denmark, Slovenia, France and Finland participating in Plato programmes. Plato Ireland (http://www.plato.ie) was established in 1995 to provide small and mid-sized enterprises (SMEs) with a unique business support forum for exploring business growth and expansion. There are currently about 1,000 participating SMEs in seven Plato regional networks around Ireland. More than 1,500 companies have already participated in Plato Ireland programmes.

In 1997 a White Paper on Human Resource Development (Department of Enterprise 1997) stated the government's view that "the primary responsibility for human resource development rests with the private sector itself ... the first approach of the state must be to facilitate the actions of business enterprises and of individuals to engage in human resource processes ... The state is prepared to back its support with adequate finance where the business sector takes a lead and responsibility." To this aim Skillnets (http://www.skillnets.com) was established in 1999. Skillnets is an enterprise-led support body whose mission is to enhance the skills of people in employment in Irish industry to support competitiveness and employability. It is funded from the National Training Fund and its stakeholders include leading employer and employee representative bodies, which include the Irish Business and Employers Confederatio, Chambers Ireland, the Construction Industry Federation, the Small Firms Association and employee/trade union representatives from the Irish Congress of Trade Unions. Skillnets is the main body in Ireland through which training networks are funded.

3.6.2 Skillnets Training Networks

Skillnets role is to act as an enterprise-led support body dedicated to the promotion and facilitation of enterprise training and other learning activities, as a key element in sustaining national competitiveness. Its mission is to provide companies and their people with new relevant and flexible methods of up-skilling staff and enhancing business performance. Skillnets strategy is to support companies from all sectors in the economy to engage in training, but does so exclusively through networks of companies under the Training Networks Programme. The network approach allows companies to achieve economies of scale, increased efficiency, lower costs and training that is more relevant to individual sectors or types of companies.

Since 1999, Skillnets has funded over 150 training networks and supported over 10,000 companies and 50,000 employees to meet their training needs. Recognizing that the enterprise-led approach has made a positive contribution to raising the profile of enterprise training in Ireland, the government granted a new five-year mandate to Skillnets for 2005–2010. Up to 55 million € was made available from the National Training Fund.

An additional 16 million € initiative which is partly funded through the European Social Fund and the National Training Fund is run on behalf of the Department of Enterprise, Trade and Employment by Skillnets Services Ltd, a wholly owned subsidiary of Skillnets. The programme began formally in January 2006 and 55 Accel (http://www.acceltraining.com) projects were funded in a wide variety of sectors and regions. Local enterprise bodies, employer and employee groups and groups of enterprises have formed training alliances which include the involvement of third level institutions in developing new skills and competency frameworks and standards as well as accredited courses and new approaches to training including e-Learning and m-Learning and other new flexible models for the delivery of in-company training.

A training network (a "Skillnet") is a group of enterprises that decide to cooperate in order to undertake a project over a sustained period that individual members of the group would be unable to undertake on their own. A typical network structure is shown in Fig. 3.11. Skillnets training networks are comprised of groups of enterprises that come together to achieve a common purpose or address a shared issue of importance to the development of skills and their long-term competitiveness. In each network, companies jointly address their training needs, collaborate with other enterprises, and engage experts, trainers, certifying bodies, industry bodies and others to work with them to achieve their goals. The core of a Skillnet is the company members. In most of the networks, a network manager is the key coordinator, providing the professional, and developmental input. Others who provide services to the network include trainers, business advisors and academics.

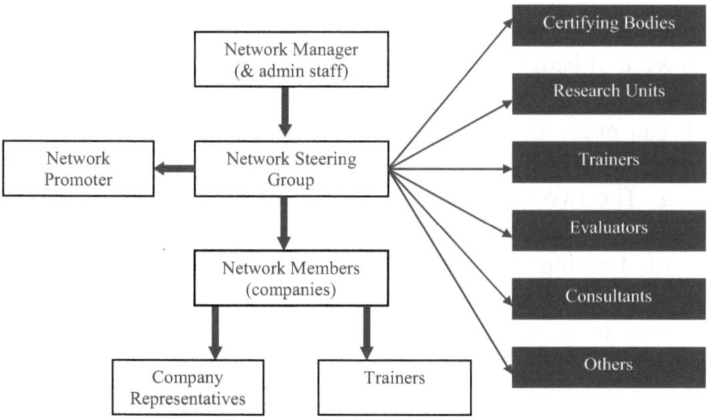

Fig. 3.11 A typical network structure

The size of companies in Skillnet and Accel training networks range from large (> 250) to medium (50 to 249) to small (10–49) to micro (< 9) (see Fig. 3.12). Skillnets places a specific priority on small companies that face particular challenges in accessing training due to their small scale and restricted resources. Eighty-two percent of companies in previous networks have less than small or micro-businesses (less than 50).

Delivering training using a network model has proved very successful over the past seven years (Skillnets 2006). There are a number of reasons for this success. The companies who actively participate in the network have more influence on the training topics, content and the even the trainers. Ninety percent of companies stated that the training as part of a network was more flexible, more accessible, more effective, better quality and lower cost in meeting their training needs. The fact that regional networks can organise training local to the companies is a major benefit. Ninety-nine percent of companies stated that the Skillnets Training Networks Programme was an effective model for the delivery of training and 98% stated they would recommend becoming a member of a training network to other companies.

Significantly for a network perspective, 70% of companies that participated in networking events organized by their network stated they were of benefit in estab-

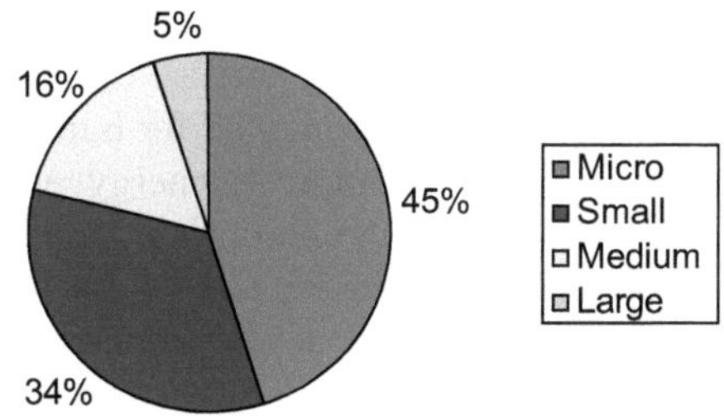

Fig. 3.12 Profile of Skillnet company participants

lishing key business contacts and generating ideas for development. Almost 90% found that these events were of benefit in sharing learning opportunities.

Skillnets do not impose a requirement regarding the formation of the network as long as the network can prove that there are benefits to be gained in terms of common training needs, cost effectiveness and value for money relative to individual company training. The two main types of networks are regional and sectoral. The former are often formed around chamber of commerce or regional business association. Often the training offered is generic in nature (e.g. management, people skills)

The majority of networks are sectoral. The training offered is often specialized to the sector. The sectoral networks tend to have less competition for the training they offer when compared to regional or local network. Sectoral networks tend to have more common training needs with other firms from within the same sector. Often companies in a sectoral network are competitors and they are often sensitive to sharing information in case they loose competitive advantage particularly at the beginning. The development of trust among the members of a network is one of the biggest barriers to attaining business synergy within a network. But as the companies train together these barriers are often lowered. Companies begin to realise that they are not direct competitors and often these "competitors" realise that their true competitors may be further away and that by sharing information with follow members in their own networks, they jointly gain a competitive advantage. There are many examples demonstrating the benefits of networks across Skillnet networks. Two case study examples are described in the following: Renewable Energy Skills Network in Sect. 3.6.3 and First Polymer Training Network in Sect. 3.6.4.

3.6.3 Case Study 1: Renewable Energy Skills

Renewable Energy Skills (http://www.renewableenergy.ie) (RES) was established in 2004 to provide training and support to plumbers, heating contractors, related trades and professionals in the emerging renewable energy heating sector. It was successful in its application for funding under the 2004/5 Skillnet Training Networks Fund. The network soon developed six core modules and provided training mainly in the mid-west region of Ireland. The network got further funding in 2006/7 and the main focus was to provide training while increasing the membership nationally, and to deliver training certified nationally.

Today the network has trained in excess of 600 individuals. The aim of the network is "to promote the use of renewable energy as a cost effective and environmentally friendly method for building heating systems, to encourage the use of technologies that improve energy efficiency; including control, insulation and build types, through appropriate training and accreditation of installers and industry professionals."

The network has developed an innovative Action on Building Energy Efficiency Programme and has successfully secured training network funding for

2008/9. From the beginning, RES has encouraged networking among trainees, promoting positive interaction and support between consumers and members. In 2006 member companies competing locally decided to jointly promote renewable energy and their expertise by firstly advertising together in a local paper and secondly organizing Renewable Energy Shows. These shows were unique on two levels. They were the first exhibitions in the country to be solely focused on renewable energy heating systems and technology and secondly they were organized by local members (plumbers and local suppliers). The shows had a number of benefits for the network. They were an excellent opportunity to attract new members (over 5% of attendees at the show registered as trade). Secondly it gave RES members an opportunity to promote themselves as experts in renewable energy heating systems. The fact that local plumbers and renewable installers could organise a show to the public, built up a spirit amongst the members and showed the magnitude of benefits that could be gained by networking. Moreover, it also demonstrated the strong demand that consumers had to discover the benefits of renewable energy systems. Five shows ran over 2006/7 attracting over 2,800 visitors. The network also hosted three national conferences, which had a number of national and international speakers.

For 2007, the network members decided to produce a 32-page *Consumer Guide for Renewable Energy*, featuring 10 articles on various aspects of renewable energy and energy efficiency written by members. For five of the 10 contributors it was their first publication. The fact that the *Consumer Guide* and the Renewable Energy Show was industry led and did not develop from a publication house or an exhibition organiser is a true testament to the power of industry networks.

3.6.4 Case Study 2: First Polymer Training

First Polymer Training (http://www.firstpolymer.com) (FPT) was set up by Plastics Ireland (formerly the Plastics Industries Association) in September 1999. This network is made of mainly medium-sized companies. The plastic sector has seen increased international competition over the past few years and companies have had to improve competitiveness to survive. This network demonstrates the resolve of an industry sector to enhance and improve the skill base through training.

FPT is run by the plastic industry and its board of directors is comprised of top-level managers from companies supplying to the construction, healthcare, electronic and automotive industries, as well as equipment and material suppliers of the plastic industry. The industry board ensures that training provision moves with the needs of the industry. Funded by Skillnets, FPT provides technical training to the Irish plastics processing industry; both in-company and at their fully equipped training centre in Athlone. The training centre has state-of-the-art plastic processing and testing equipment and high-quality conferencing facilities.

The overall objectives of the network are to support companies in the plastic processing sector, primarily those who are aiming to improve efficiencies to combat the threat of low-cost parts from abroad, and also those companies wishing to move to higher-value products. FPT continues to develop and deliver training courses in areas as diverse as injection moulding, extrusion, blow moulding, mould design, plastic part design, plastic materials, etc. FPT also provide courses which support the plastic manufacturing process, but are equally applicable to any manufacturing industry, such as statistical process control (SPC), design of experiments (DOE), electrical, pneumatic and hydraulic maintenance, and a range of "lean manufacturing" courses including 5S, value stream mapping, and so on.

FPT has developed innovative methods of training; analogies and computer simulation programs are used to allow participants to visualise the processes under discussion. All FPT courses are very hands-on. Most courses are 70% practical and no more than 30% theory, which FPT believe to be the most effective way to deliver technical training.

During 2005 FPT began promoting efficient manufacturing techniques and tools (called lean manufacturing) to their network members. In Early 2006, FPT successfully secured the support of Accel for The Kaizen Project (http://www.kaizenproject.ie). The Kaizen Project is a focused network of manufacturing companies working to implement Kaizen continuous improvement tools and techniques, and build in-company lean expertise. The network is made up of 10 core companies who are undergoing a lean transformation programme over a two-year period, and a wider network of 20 companies who are collaborating and learning from the project. In early 2007 an additional five companies opted to undergo the lean transformation for the second half of the project.

The main aims of the program are to improve competitiveness of the companies and prove that a network approach will provide the momentum to implement a continuous improvement philosophy within a company. By working as a network, elements of peer pressure and competition helps focus the companies: the groups set monthly targets and milestones for each company to meet. Each company employs a full-time task manager devoted entirely to the implementation of the training project. The project manager and task manager, as well as the trainer(s), work closely with each company's own task managers to ensure the success of the group.

As of March 2007 The Kaizen Project has trained over 1,000 participants in over 10 companies. All these companies have participated in running Kaizen pilot projects. Six of the 10 companies have increased their participation in 2007 with many of them running projects company-wide. The improvements within the organizations are impressive. One company increased output per head by 53%. Another company doubled production output in the same production time period.

3.6.5 Some Concluding Remarks

Training is an important activity for companies to ensure their competitiveness. Training networks, rather than individual company training, is an excellent method to implement training programmes, especially for SMEs and micro-companies. The Irish government who have provided substantial funding through the Skillnet and more recently the Accel programmes have recognized this. There, since a decade, the Skillnet and Accel programmes support a wide range of industrial training needs through regional and sectoral networks.

3.7 SME Networks for Car Recycling in Hungary: an Agent-Based Approach

G. L. Kovács and G. Haidegger – Hungarian Academy of Sciences

SME networks have different characteristics across Europe. In Hungary and in some other similar-sized countries the distribution of companies varies from near-by agglomerations around one country up to Europe-wide linkages. Although the clusters are rather specialized, a relatively high number of different skill and knowledge fields are employed. Hungarian clusters typically consist of rather small SMEs with 5–50 persons. Special organizational structures do not exist, whereas there are leading firms and common marketing activities for the majority of Hungarian clusters.

Recycling and reuse of end-of-life vehicles (ELV) is one of the most challenging issues of the century to sustain development, to decrease pollution and to decrease energy and material consumption. The main roles in this field are given to the several hundreds (thousands) of SMEs. The recent E-Mult project (E-Mult 2005) suggests a systematic approach to solve these problems in the form of ICT solutions using networked, multi-agent systems with advanced knowledge management supports (Kovács, Haidegger 2001).

The E-Mult project goal is to achieve multi-threaded, agent-based multi-regional and transnational networks of SMEs working together in the ELV recycling industry in Europe based on highly dynamic business models. As a result it provides innovative solutions in the form of a set of free SW building blocks (BB) of a highly scalable, open architecture, an agent-based platform for operation of dynamic networks of SMEs, including powerful distributed decision support systems for network management and an infrastructure for knowledge sharing.

3.7.1 The E-Mult Project Approach

In order to satisfy the requirements of the European Waste Management concerning ELVs, the participating SMEs must have supporting tools, a methodology to efficiently run highly sophisticated, complex operations in a strongly interrelated (networked) manner.

To have a methodology and platform to support SMEs organizing dynamic co-operative networks for market-driven ELV recycling, our approach is to combine agent-based technology with adaptive, rule-based reasoning and statistical methods. Based on the thorough analysis of needs of SMEs and current systems for network management, it has been considered that the only way to meet the strategic objectives of SMEs is to obtain relevant solutions in the following main fields:

- An innovative autonomous, multi-layer, multi-agent platform as a highly flexible and reliable technology for distributed decision support and knowledge sharing, low-cost and acceptable by SMEs.
- A set of innovative procedures and algorithms for effective network management based on the combination of probabilistic reasoning, product/process models and statistical methods solving three key identified problems in management of dynamic networks of SMEs:
  - New procedure(s) for optimal sharing of resources and distribution of activities within and between different networks.
  - Adaptive algorithms for real-time decision making on dismantling strategy and scheduling of the shop floor operations to solve the above mentioned problems regarding uncertainty in input material (ELV), high diversity, etc.
  - A new method for tracing and measuring added value along the recycling chains solving the harmonization between added value in multi-threaded, "backward" (collection, dismantling) processes and "standard" added value within forward processes in production of new products based on recyclates.
- An innovative methodology for identification of optimal enterprise models of multi-threaded, dynamic networks of SMEs, where both economic, ecological and legislative aspects, and complex interactions between processes and market needs have to be taken into account.
- Methods and tools for practicable knowledge management (KM), combining different approaches (heuristic reasoning and models) and implemented using advanced agent technology, including new forms for representation of the experience-based knowledge. A special challenge is to establish an adequate ontology approach (assuring easy maintenance of ontology) applicable in multi-language and multi-cultural networks of recycling SMEs.

The main innovations of the developed system are holistic and highly harmonized solutions in the above fields. The project offers a strategy with extreme flexibility (supported by the methodology and tools) in the recycling sector to attack problematic areas, which have not yet been solved by SMEs in this business sector.

1. Practical goals

In order to provide a holistic solution to the above challenges some operational goals have been defined, which result in the development of:

- A set of software (SW) BBs of an agent-based platform for different services, supporting SMEs to efficiently set up and operate dynamic networks (decision support services and a set of services for knowledge sharing, etc.).
- Methodology adapted to SMEs to easily establish and manage innovative dynamic networks and set up and operate an ICT infrastructure for e-collaboration, including
  - Enterprise/business models of SME-driven networks in ELV recycling domain.
  - Methods and tools to set up and operate dynamic networks of SMEs.
  - Methods for management of dynamic networks (including organizational, cultural and legislative aspects in transnational networking).
  - Method for setting up (customized) ICT infrastructure for an advanced knowledge sharing environment.
  - Ontology applicable for ELV recycling SMEs.
- Two business cases (BC1 and BC2) in six countries, i.e. the establishment of twp transnational networks of ELV recycling SMEs, serving as a baseline for the project, enabling the verification and demonstration of the developed methodology and SW tools. Both the vertically networked structures (BC1, networking of car dismantlers targeting a provision of recycled material to recyclates users) and the horizontally networked structures (BC2, improved procurement of dismantled parts and materials) will be demonstrated.
- Web-based Knowledge Forum where SMEs are able to meet other SMEs, ICT providers and RTD partners.

The brief overview of the state-of-the-art in two key problem areas, relevant for the project is provided below.

2. Agent-based technology: state-of-the-art

It is identified as the optimal approach to provide an ICT platform for communication and management of dynamic SME networks. Although intelligent SW agents have been known for about 30 years, they are only now becoming more important, thanks to the growth of the Internet. The most relevant aspect is integration of highly intelligent agents in a multi-agent system, running in a distributed environment.

Multi-agent-based technology is used for distributed planning, auction and learning mechanisms, etc., in highly dynamic networks. There are still many issues and unsolved problems for which additional research is needed (Krupansky 2004). Recently none of the existing solutions reaches the required level of sophistication, flexibility and robustness needed by the ELV recycling SMEs.

3. Dynamic networks of SMEs: state-of-the-art

SME networks are established to achieve the “collaborative advantage” (Kanter 1994) to be realised by reduction of costs, better exploitation of resources, opening of markets, enforcement of standards and mitigation of risks. Generally speaking, a type of network can be distinguished by its control, by either hierarchical or flat organization (Winand 1998) and its stability. We shall focus on the modelling of dynamic networks, searching for new solutions where interconnections and interoperability among a number of networks may play a crucial role. The state-of-the-art Internet-worked economy solutions applied as support for different forms of smart networking, including SCM systems, distributed planning systems, Extended ERP or ERP II Systems, etc., are mainly oriented to large system applications (SAP, Baan, Siebel), differing in complexity and price. These solutions are not applicable for recycling SMEs, where we need rather sophisticated solutions to achieve extreme flexibility in processes (Terk, Haidegger et al. 2005).

3.7.2 Two Project Test Sites

There will be two experimental clusters/business cases to show common (and different) results of the cooperation within the project:

The BC1 environment will serve for development of a show case of technology to facilitate the effective validation and efficient demonstration of the potential business benefits envisaged by the project. BC1 will demonstrate the full prototype of the international vertical dynamic networking aiming at provision of recycled materials for the new products in day-to-day operation with all BC1 SMEs. This will incorporate further tests and measurements within the BC1, providing major input for the final dissemination activities and for the development of the final exploitation strategy.

Within the BC2 environment the demonstrator as a show case of technology will be developed to facilitate the effective validation and efficient demonstration of the potential business benefits envisaged in this business case by the E-Mult project. BC2 will demonstrate the full prototype of the international horizontal dynamic networking aiming at provision of recycled materials for the new products in day-to-day operation with all BC2 SMEs. This will again incorporate further tests and measurements within the BC2, providing major input for the final dissemination activities and for the development of the final exploitation strategy, as done for the BC1 environment.

BC2 will contain all the important features of the cooperation, i.e. basically a Web-based supply and demand chain system. The number of SMEs participating in the clusters is only four or five in the beginning. There is one SME from Austria working together with three to four Hungarian SMEs. Geographically there is a maximum of 3 to 400 kilometres between any two participants. This makes the application of agent-based, networked system solutions really applicable. The number of participants will surely increase due to success demonstrated during the first application period.

A very simplified, practical view of cooperation of SMEs as a demand and supply network is that every time a new car arrives for dismounting all internal and external requests from all partners and some forecasts are taken into account when decision is made on re-using a part or not. On the other hand any time a request arrives to any of the partners a decision can be made on what (and by whom) to offer to the potential customer. Both types of the above-mentioned decisions are supported by intelligent/expert programming tools on one hand; on the other hand the Knowledge Forum is always available and ready to assist in decision making in such real and simulated situations. BC2 will serve as an international test-bed to validate the effectiveness of the project results:

The harmonization of selling scrap metal is also beneficial for the SMEs joining the network: they can negotiate for better prices with the large scrap material handlers.

3.7.3 Governance and Managerial Mechanisms

Due to the character of the Web technologies and the intelligent, agent-based networks in the system there are no special governance structures, nor specific managerial mechanisms. All partners have the same rights, which basically means that they can set requests and ask questions to any/all partners at any time. At the same time any partner can achieve all available information at any time. Naturally, as we will show later there are different access rights in the Knowledge Portal/Forum of the system. All firms can keep their privacy at the necessary level (Haidegger, Kovács 2003).

3.7.4 Some Details on the E-Mult Project

In the following sections some important aspects of our ELV recycling SME network will be discussed.

1. Business applications for SMEs in ELV recycling

The existing business applications for SMEs in ELV recycling are just partly covering their needs. The few typical examples are: IDIS, Mitan Autoverwertung, Combera-AUTOREC, ETA/AutoRecy, CARCYCLE, Carparts-uk.com, the UK's e-commerce site, and Callparts, German e-shop offering parts selling, orders and delivery information via the Internet. The SW of the company Althaler and SW ARES of the company Ambit, support the recycling process through ELVs and parts recording and stocking, dismantled parts selling in the shop via call centres and central server (WebParts), disposal management according to ELV directive RL 2000/53/EG, etc. Ferguson (2002) explains the key research problems in detail, indicating the above-mentioned main needs.

2. Knowledge management methods and tools and ontology

Knowledge sharing within multi-threaded networks is related to a number of fundamental problems such as human acceptance and motivation issues, ontology problems, correlation of different types of knowledge, treatment of experience-based and often incomplete and/or ill-structured knowledge, etc. Management of tacit knowledge including its capturing, maintenance and sharing is still not efficiently solved in industrial practice. The typical problem is to ensure maintenance and update of knowledge systems by SME staff, missing it leads to a relatively short lifetime of KM systems in manufacturing companies, i.e. knowledge becomes obsolete. Existing, often powerful methods and ICT tools, do not satisfy many of these requirements. E-Mult (2005) applies and enhances advanced, content-sensitive knowledge management (KM) methods and tools.

Ontologies are often considered to be the most powerful means to solve the problem of efficiently storing, retrieving and sharing of knowledge within complex networks. A number of specific ontologies and tools to set up/maintain KMs have been or are currently developed. However, the ontologies to support knowledge sharing within the recycling area are still in development. Especially needed is a means for continuous update/maintenance of ontologies enabling long life of KM systems. To reuse or extend ontologies that can be applied in ELV recycling context is planned in the near future, as well as to provide innovative ways to update ontology applied within the networks of SMEs.

3. The target system

The project generates an agent-based ICT platform, incorporating a set of SW building blocks (BB), which enables SMEs to set up and operate their own dynamic networks (system for dynamic network management) and to establish a Knowledge Forum as a means for an advanced knowledge sharing.

4. Dynamic network management

The ICT platform provides a means for communication within dynamic SME networks, but its main purpose is to provide decision support covering the basic problems within such networks (see Fig. 3.13; IAG meaning the Industrial Associations Group):

- Sharing of resources and distribution of activities among SMEs
- Decision support on dismantling strategy
- Scheduling and added value tracking

Using the SW BBs, SMEs are able to efficiently build their own (customized) platform and include functionalities they need to operate the networks. The project develops critical functionalities, but the platform may include other standard functionalities to manage such networks (logistics, stock handling, Internet shop for spare parts, quality assurance, etc.), which are already covered by the available market business applications for recycling SMEs.

5. Knowledge sharing infrastructure: E-Mult Knowledge Forum

The Knowledge Forum includes a powerful EU-wide database, and it is a place where SMEs are able to identify potential innovative applications of recyclates, to

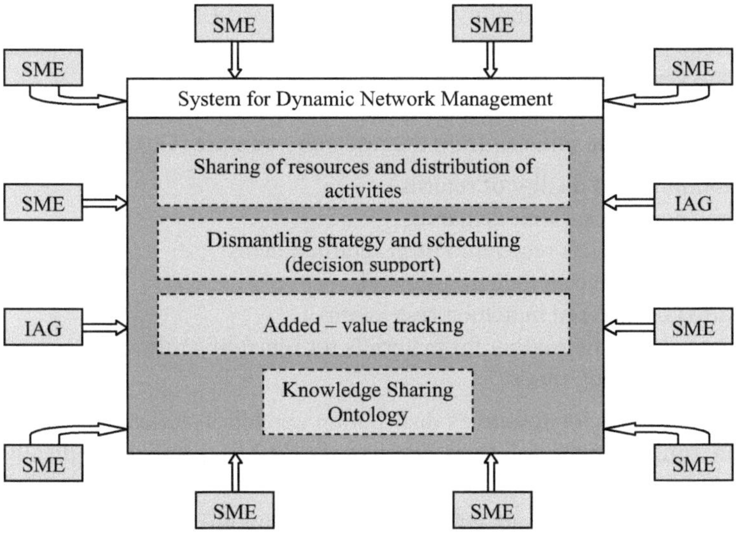

Fig. 3.13 Main functionalities of dynamic network management

identify processes needed for such applications, to identify how to set up value chains (using methodology and tool developed), to identify the most suitable partners, establish common and *members only* areas for knowledge sharing within networks, and to "meet" RTD partners and their ideas, etc.

On top of that, the system provides all relevant knowledge related to legislative and standardization issues, etc. All partners are involved in the establishment of the knowledge-sharing infrastructure within the Knowledge Forum, aiming to build knowledge communities of recycling SMEs.

6. General requirements

The following main requirements should be met by the resulting system:

- Low cost for the involved SMEs, easy to use and learn
- Application of local languages, easy customization (and integration)
- Efficient support of communication/knowledge sharing among SMEs
- Support to more open aspects of collaboration between SMEs
- Record the key information associated with all actors, assuring traceability
- Assure common terminology/ontology and safety/security
- Support mobile users, and so on

A set of SW BBs has been developed in order to enable easy re-configuration and customization of the systems, meeting diverse specific needs of SMEs in this sector. To meet all requirements a strong production planning and the optimal, joint usage of all resources has to be solved. The objective is to provide functionality to cover different time horizons depending on specific network needs (hourly, daily, weekly, monthly). The main problems are the multi-threaded character of networks, the low controllability and predictability of ELV quantity, state, models in different plants, and difficulty in the prediction of market demands which may change very abruptly.

A number of characteristics that significantly complicate production planning and control activities may be inferred from the research literature (Guide 2000):

- The uncertain timing and quality of returns
- The need to balance returns with demands
- The uncertainty of materials recovered from returned items
- The requirement for a reverse logistics network
- The complication of material matching restrictions
- The problems of stochastic routing for materials for remanufacturing
- Highly variable processing times

The classical algorithms for resources distribution applied in Extended ERP II (and included in some current solutions for recycling SMEs) will be radically improved by including statistical forecast (of ELVs and of the market), probabilistic, rule-based reasoning (RBR), in combination with models of networks and models of products/services and agent-based deployment of algorithms. Agents support "negotiation" among the networks to find an optimal distribution of work and perfectly support distributed decision making within networks.

7. Algorithm(s) for real-time decision making and scheduling

These algorithms are developed based on a combination of classical algorithmic and statistical forecast approaches and probabilistic RBR. The functionality to be provided here has to support local management of the dismantling process within the networks. As the costs of ELV dismantling strongly depend on the depth of dismantling, and as the achievable market prices of spare parts and re-usable material are relatively low, due to intensive manual work the proper dismantling of some parts may be rather expensive, it is necessary to carefully study for each part whether it is reasonable to dismantle it completely or is it better to send the ELV straight to the shredder.

Dismantling SMEs require very accurate planning for each ELV. The depth of dismantling (dismantling strategy) has to be decided on-the-fly, for each specific ELV based on considerations of the following: costs required for car dismantling depending on car type and its state, prices and actual demands on the market for spare parts and re-usable material, cost of re-usable parts and material stocking and current level of parts in inventory, transport costs, (hazardous waste) material which has to be removed from the car due to ecological reasons (legislative aspects).

All the above-mentioned problems related to uncertainty in input material, and multiple, dynamically changing networks requirements (including market demands) play a crucial role in the decision-making process. On top of this, the problem of the dismantling strategy definition is also the low process standardization, causing the difficulties to predict time and efforts/costs for different processes (taking into account extreme diversity of models/state of ELVs to be dismantled, requesting high flexibility of all processes). Once the dismantling strategy is defined, the scheduling of work has to be carried out based on instant cost-benefit estimation.

8. Multi-agent system (MAS)

A multi-agent system is part of the E-Mult system, which serves as monitoring middleware between E-Mult and external systems. We tailor the monitoring agents for use with particular sites. The resulting benefits are the transparency of external data sources, the use of a hierarchy composed of both management and monitoring agents instead of only monitoring agents. The monitoring tasks/ assessment patterns can be dynamically specified within the proposed frame, and finally run-time creation of task agents helps to keep the system compact and prevents useless consuming of resources.

The main issues of MAS to be solved are the following:

- Data handled by the MAS and users involved in the system
- Set-up parameter definition and data flow for monitoring tasks

3.7.5 A Few Remarks to Conclude

A new, more effective cooperation/collaboration network and environment will be offered for European SMEs, who are active in ELV dismantling, reuse and recycling. The network will work as a cooperative demand and supply network, and will be realised as a multi-threaded, multi-agent system due to the efforts of a consortium consisting of Industrial Associations Groups (IAG), SMEs and academic partners from several European countries (Haidegger, Kovács, Bàmer 2003).

References

Akao Y (1990) Quality Function Deployment: Integrating Customer Requirements into Product Design. Productivity, Cambridge, MA

Anand B, Khanna T (2000) Do firms learn to create value? The case of alliances. Strategic Manage J 21(3):295–315

Andersson et al. (2004) Relationship dynamics: developing business relationships and creating value. In: Mahnke V, Pedersen T (eds) Knowledge flows, governance and the multinational enterprise. Macmillian, New York

Antonelli D, Villa A (2007) Network analysis by the CODESNET Approach. In: Establishing the Foundation of Collaborative Networks. Springer, Boston. pp. 67–74, Vol. 243/2007, ISBN: 978-0-387-73797, DOI: 10.1007/978-0-387-73798-0

Appley DG, Winder AE (1977) An Evolving Definition of Collaboration and Some Implications for the World of Work. J Appl Behavioral Sci 13(3):279–291

Arda Y, Hennet JC (2006) Inventory control in a multi-supplier system. Int J Prod Econ 104(2):249–259

Arda Y, Hennet JC (2007) Inventory control in a decentralized two-stage make-to-stock queuing system. Special issue on Production Coordination and Inventory Policies. Int J Syst Sci (in press)

Asheim BT, Isaksen A (2002) Regional innovation systems: the integration of local 'sticky' and global 'ubiquitous' knowledge. J Technol Transf 27(1):77–86

Babich V, Ritchken PH, Burnetas AN (2007) Competition and Diversification Effects in Supply Chains with Supplier Default Risk. Manufact Service Oper Manage 9(2):123–146

Barad M, Gien D (2001) Linking improvement models to manufacturing strategies – a methodology for SMEs and other enterprises. Int J Prod Res 39(12):2675–2695

Barley S, Freeman J, Hybels R (1992) Strategic alliances in commercial biotechnology. In: Nohria N, Eccles R (eds) Networks and Organizations: Structure, Form and Action. Harvard University Press, Cambridge, MA

Batt PJ, Purchase S (2004) Managing collaboration within networks and relationships. Indust Market Manage 33(3):169–174

Becattini G (1990) Industrial district and interfirm cooperation in Italy: the marshallian industrial district and the socio-economic notion. International Institute for Labour Studies, Geneva, pp. 37–51

Bergman EM, Feser EJ (1999) Industrial and Regional Clusters: Concepts and Comparative Applications. Regional Research Institute, West Virginia University, Morgantown, WV

Biswas S, Roy S, Seshagiri S (2007) Collaboration in Indian SME Clusters: A Case Study. Paper presented at the 3rd International Conference on Communities and Technologies, 28–30 June 2007

Bititci US, Martinez V, Albores P, Parung J (2004) Creating and managing value in collaborative networks. Int J Phys Distrib Logistics Manage 34(3/4):251–268

Bitran GR, AC Hax (1977) On the Design of Hierarchical Production Planning Systems. Decision Sci 8(1):28–55

Bossert J (1991) Quality Function Deployment: A Practitioner's Approach. ASQC Quality, Milwaukee, WI

Burt R (1992) Structural Holes: The Social Structure of Competition. Harvard University Press, Cambridge, MA

Cachon GP, Netessine S (2004) Game theory in Supply Chain Analysis. In: Simchi-Levi D, Wu SD, Shen ZJ (eds) Handbook of Quantitative Supply Chain Analysis: Modeling in the eBusiness Era. Kluwer, Dordrecht

Cachon GP, Zipkin PH (1999) Competitive and cooperative inventory policies in a two-stage supply chain. Manage Sci 45(7):936–953

Chopra S, Meindl P (2004) Supply Chain Management – Strategy, Planning and Operation, Pearson, 2nd edn. Prentice Hall, New York

Clancy P, Twomey M (1997) The Irish Popular Music Industry: An Application of Porter's Cluster Analysis. Research Series Paper No. 2. National Economic and Social Council, Dublin

Culliton J (1992) A Time for Change: Industrial Policy for the 1990s. Report of the Industrial Policy Review Group. Stationery Office, Dublin

De Kok AG, Graves SC (eds) (2003) Supply Chain Management: Design, Coordination and Operation. Elsevier, Oxford

DeBresson C, Amesse F (1991) Networks of innovators: a review and introduction to the issue. Regional Policy 20(5):363–379

Delapierre M, Mytelka L (1998) Blurring boundaries: new inter-firm relationships and the emergence of networked, knowledge-based oligopolies. In: Colombo MG (ed) The Changing Boundaries of the Firm: Explaining evolving inter-firm relations. Routledge, London

Department of Enterprise and Employment (1997) Human Resource Development, White Paper. 1997. Ireland Government Publications, Dublin

E-Mult (2005) European Multi-threaded Dynamic SME Networks for Market-Driven ELV Recycling. Contract EC-IST- 27212/, Annex I. Description of Work. PIAP, Warsaw, pp. 1–101. http://195.187.100.149/emult/. Accessed 2008

Ferguson NS (2002) An information system for end of life product recovery. Dissertation,

CIMRU, National University of Ireland, Galway
Gadde LE, Huemer L, Hakansson H (2003) Strategizing in industrial networks. Indust Market Manage 32(5):357–364
Gavirneni S, Kapuscinski R, Tayur S (1999) Value of information in capacitated supply chains. Manage Sci 45(1):16–24
GFP (2005) Enabling SMEs to enter the International supply chain. The World Bank Group, GFP Exploratory Notes, June 2005. http://www.gfptt.org. Accessed 2008
Granovetter M (1985) Economic action and social structure: the problem of embeddedness. Am J Sociol 91(3):481–510
Griffin A, Hauser JR (1993) The Voice of the Customer. Market Sci 12:1–27
Guide DV (2000) Production planning and control for remanufacturing: industry practice and research needs. J Oper Manage 18(4):467–483
Gulati R, Nohria N, Zaheer A (2000) Strategic Networks. Strategic Manage J 21(3):203–215
Haidegger G, Kovács Gy (2003) On general issues of Technology Transfer in Hungary. In: Kovács Gy, Haidegger G (eds) EW-ISME Success Stories Seminar 20 Oct 2003. MTA SZTAKI, Hungary
Haidegger G, Kovács Gy, Bámer B (2003) Recycling Technology Implementation by the Integration of SME Operations. In: Tóth T, Bikfalvi P, Göndri Nagy J (eds) WESIC 2003, Advanced Technologies in Manufacturing 28–30 May, University of Miskolc, Proceedings of the 4th Workshop on European Scientific and Industrial Collaboration vol I, pp. 611–615
Hakansson H (1990) Technological collaboration in industrial networks. Eng Manage 8(3):371–379
Hakansson H, Ford D (2002) How should companies interact in business networks. J Bus Res 55(2):133–139
Hardy C, Phillips N (1998) Strategies of engagement: lessons from the critical examination of collaboration and conflict in an interorganizational domain. Organ Sci 9(2):217–230
Harland CM, Lamming RC, Zheng J, Johnsen TE (2001) A taxonomy of Supply Networks. J Supply Chain Manage (Fall 2001):21–27
Himmelman AT (1992) Communities Working Collaboratively for a Change. The Himmelman Consulting Group, Minneapolis
Huxham C (1996) Collaboration and collaborative advantage. In: Huxham C (ed) Creating Collaborative Advantage. SAGE, Thousand Oaks, CA
Kale P, Singh H, Perlmutter H (2000) Learning and protection of proprietary assets in strategic alliances: building relational capital. Strategic Manage J 21(3):217–237
Kanter RM (1994) Collaborative advantage: the art of alliances. Harvard Bus Rev 72(4):16–108
King R (1995) Designing Products and Services that Customer Wants. Productivity, Portland
Knights D, Murray F, Willmott H (1993) Networking as knowledge work: a study of strategic interorganizational development in the financial services industry. J Manage Studies 30(6):975–995
Kovács GL, Haidegger G (eds) (2001) Business Growth through European Partnerships, Conference for SME-s. Proceedings of the EU IPS-2001-42012 EW ISME project's Final Conference. MTA SZTAKI, Hungary
Krupansky J (2004) Agtivity: State of the Art for Software Agent Technology. http://agtivity.com. Accessed 2008
Lariviere MA, Porteus EL (2001) Selling to the newsvendor: an analysis of price-only contracts. Manuf Service Oper Manage 3(4):293–305
Lewis DJ (1990) Partnership for Profit: Structuring and Maintaining Strategic Alliances. Free Press, New York
Lincoln JR, Gerlach ML, Takahashi P (1992) Keiretsu networks in the Japanese economy: a dyad analysis of intercorporate ties. Am Sociol Rev 57(5):561–585
Martin R, Sunley P (2003) Deconstructing clusters: chaotic concepts or policy panacea? J Econ Geogr 3(1):5–35
Myhr N, Spekman RE (2005) Collaborative supply-chain partnerships built upon trust and electronically mediated exchange. J Bus Indust Market 20(4–5):179–186

Nix N, Zacharia ZG, Lusch RF, Bridges WR, Thomas A (2004) Keys to Effective Supply Chain Collaboration: A Special Report from the Collaborative Practices Research Program. The M. J. Neeley School of Business, Texas Christian University, Fort Worth

O'Connell L, Van Egeraat C, Enright P (1997) The Irish Dairy Processing Industry: An Application of Porter's Cluster Analysis, Research Series Paper No. 1. National Economic and Social Council, Dublin

O'Gorman C, O'Malley E, Mooney J (1997) The Irish Indigenous Software Industry: An Application of Porter's Cluster Analysis, Research Series Paper No. 3. National Economic and Social Council, Dublin

Osborne MJ, Rubinstein A (1994) A Course in Game Theory. MIT Press, Cambridge, MA

Pander S, Wagner R (2005) Unternehmensübergreifende Zusammenarbeit in der Automobilentwicklung – durch erfahrungsgeleitete Kooperation die Grenzen der Planbarkeit überwinden. Hampp, Mering

Phillips N, Lawrence TB, Hardy C (2000) Interorganizational collaboration and the dynamics of institutional fields. J Manage Studies 37(1):23–45

Porter ME (1990) The Competitive Advantage of Nations. Free Press, New York

Pouly M, Monnier F, Bertschi D (2005) Success and Failure Factors of Collaborative Networks of SME. In: Camarinha-Matos LM, Afsarmanesh H (eds) Collaborative Networks and Their Breeding Environments, vol 186. Springer, Berlin Heidelberg New York

Powell WW, Koput KW, Smith-Doerr L (1996) Interorganizational collaboration and the locus of innovation: networks of learning in biotechnology. Admin Sci Q 41(1):116–146

Radhakrishnan S, Srinidhi B (2005) Sharing Demand Information in a Value Chain: Implication for Pricing and Profitability. Rev Quant Finance Account 24(1):23–45

Ringle CM (2004) Kooperation in virtuellen Unternehmungen – Auswirkungen auf die strategischen Erfolgsfaktoren der Partnerunternehmen. Gabler, Wiesbaden

Rowley T, Behrens D, Krackhardt D (2000) Redundant governance structures: an analysis of structural and relational embeddedness in the steel and semiconductor industries. Strategic Manage J 21(3):369–386

Sabel CF (1996) A measure of federalism: assessing manufacturing technology centers. Regional Policy 25(2):281–307

Science Technology and Innovation Advisory Council (1995) Making Knowledge Work For Us. Stationary Office, Dublin

Shapiro J (2001) Modeling the Supply Chain. Brooks/Cole Thomson Learning, Pacific Grove, CA

Shoham Y (1993) Agent-oriented Programming. Artif Intell 60:51–92

Simchi-Levi D, Kaminsky P, Simchi-Levi E (2000) Designing and managing SC. McGraw Hill, Boston

Skillnets (2006) The Skillnets Company Satisfaction Survey 2006. http://skillnets.ie/skillnets/resources/publications.html. Accessed 20 Dec 2007

Slack N (1994) The importance performance matrix as a determinant of improvement priority. Int J Oper Product Manage 14(5):59–75

Subramaniam V (1998) Efficient sourcing and debt financing in imperfect product markets. Manage Sci 44(9):1167–1178

Sucky E (2007) A model for dynamic strategic vendor selection. Comput Oper Res 34:3638–3651

Tayur S, Ganeshan R, Magazine M (eds) (1999) Quantitative Models for Supply Chain Management. Kluwer Academic, Dordrecht

Terk E, Haidegger G et al. (2005) Enterprises in technology-intensive business. Toolkit for coping with international environment and developing management competences. Module 2: East-West cooperation. Estonian Institute for Futures Studies, Tallinn, pp. 26–39

Tomkins C (2001) Interdependencies, trust and information in relationships, alliances and networks. Account Organ Soc 26(2):161–191

Van Brussel H, Wyns J, Valckenaers P, Bongaerts L, Peeters P (1998) Reference architecture for holonic manufacturing systems: PROSA. Comput Indust 37:255–274

Villa A, Bellomo D, Cassarono I (2005) Uncertain demand and supply network management: application to a regional health care service. In: 16th IFAC World Congress, Prague

Villa A, Cassarino I (2005) CODESNET: an approach to modelling COllaborative Demand and Supply NETwork. In: Collaborative Networks and their Breeding Environments. Springer, Berlin Heidelberg New York

Von Hippel E (1988) Sources of Innovation. Oxford University Press, New York

Wasserman S, Galaskiewicz J (1994) Advances in Social Network Analysis: Research in the Social and Behavioral Science. SAGE, Thousand Oaks, CA

Wasserman S, Faust K (1994) Social Network Analysis: Methods and Applications. Cambridge University Press, Cambridge

Wikipedia: The Free Encyclopedia (2008) Graph theory. http://en.wikipedia.org/wiki/Graph_theory. Accessed 2008

Winand U (ed), Klaus N (1998) Unternehmensnetzwerke und virtuelle Organisation. Schäffer-Poeschel, Stuttgart

Wood DJ, Gray B (1991) Toward a Comprehensive Theory of Collaboration. J Appl Behav Sci 27(2):139–162

Zeller C (2002) Project teams as means of restructuring and development in the pharmaceutical industry. Regional Studies 36(3):275–289

Chapter 4
The CODESNET Approach to SME Networks Analysis

Abstract The overview of SME clusters and networks in some European countries, and the description of approaches and actions to promote the emergence of new aggregations – presented in Chap. 3 – include also the main concepts and ideas which originated from the EU-funded CODESNET project. Since the definition of the project proposal, the scope was not to produce a collection of data by which to generate statistics, but to create a new conceptual model of a SME aggregation, sufficiently general to be used to describe a supply chain as well as a complex multistage line. Also, the descriptive model investigated, is mostly oriented to be a tool for describing and evaluating the performance of the underlying networked industrial body. Therefore, the scope of Chap. 4 is to illustrate the model generation as well as the performance analysis methodology, which has been designed on the basis of the model itself. There are two complementary goals: on one side, to generate a performance evaluation method to *compare* two different industrial systems, such to identify which one presents the strongest features; on the other hand, the same method should help the analyst to *understand why* an industrial system appears to be effectively stronger than others. Then, such a methodology needs to be supported by a catalogue of SME networks presentations, to allow comparisons, and by an archive of technical reports, from which to derive concepts and suggestions to understand and justify either the weakness or the strength of a network.

4.1 A Model to Represent SME Networks: Structure, Governance and Interactions with Markets

A. Villa and D. Antonelli – Politecnico di Torino

The description of a SME network requires the formulation of a conceptual model of the network itself, including a comprehensive representation of the structure, the organization of interactions among the component enterprises, and the connections between the network and its socio-economic context. This model should be the basis of a performance evaluation methodology, which utilizes – for estimating the efficiency, effectiveness and convenience of any networked industrial body – a collection of industrial cases (i.e. presentations of existing networks) and technical and scientific models and procedures (selected from existing literature).

The network model developed in the CODESNET project is based on the basic idea that any SME aggregation can be analysed according to three main complementary viewpoints.

The first viewpoint concerns the analysis of the production and logistic structure, which connects the enterprises, to be performed by considering: (a) how production operations are attributed to the different SMEs; (b) which type of logistic organization is used; (c) what are the production capacities of the SMEs. These evaluations will allow the analyst to assess how the network organization could improve the production volumes of the different SMEs, the amount of personnel employed there, and the transport capacities over the internal logistic network.

As a second viewpoint, refers to the governance organization (i.e. the management responsibility attributed to each SME and the amount of information that each SME has at its own disposal for management purposes), the analyst is asked to consider the types of internal agreements and control mechanisms, and the types of agreements with external bodies, to estimate the characters of the organization chart at the network level, the functionality of the coordination committee, if any, and the coordination strategies decided and implemented there.

Then, the third viewpoint refers to the analysis of the network interactions with the external environment, meaning the analysis of the commercial agreements with clients/suppliers, the kinds of strategies to negotiate at the network level, both production resources and labour, the types of policies to plan, again at the network level, innovation programs, in order to estimate the dynamic evolutions of the market penetration, the labour employment, and the risk-capital acquisitions.

In formal terms, these considerations give rise to a "meta-model" of a SME network, which is a conceptual representation of the main components of a network and their connections and interactions (see Fig. 4.1). Each component is described in terms of its main functionality by which it contributes to the network behaviour; each component's input refers to information and prescriptions driving the component's functionality and each output concerns a result of the component's activity. In practice, the meta-model includes:

- The network's *operation structure* (OS), representing the graph of the logistic connections among the firms
- The network's *organization arrangement* (OA), referring to the SME network governance and the information pattern which links the firms together
- The network's *interactions with the socio-economic environment* (ISEE), referring to the negotiations with the external markets with which the SME network itself operates

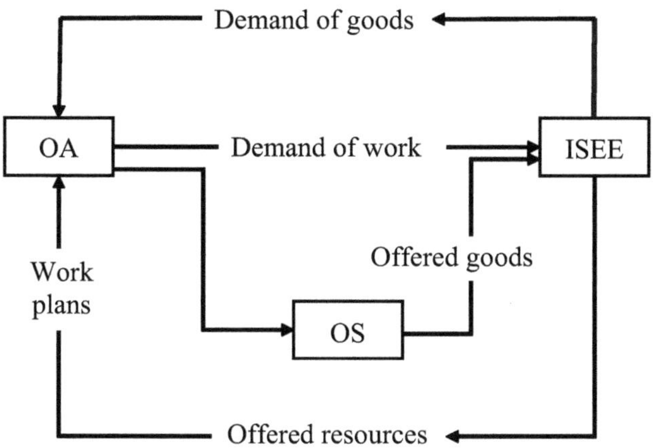

Fig. 4.1 The conceptual model of a SME network

This meta-model and its components have been inferred from industrial evidence.

The *operation structure* refers to the graph of interactions linking the enterprises together, through flows of parts, information, controls, money, and so on. Each node of this graph is an autonomous enterprise, and plays the role of an individual decision-maker (DM).

The *organization arrangement* concerns the over-firms organization devoted to managing cooperation of the enterprises. In principle, its scope is to harmonize the production and service plans of the different enterprises such that the delivery of final products to the output markets matches the demand.

The *interactions with the socio-economic environment* refers to the output interface towards external markets. Note that negotiations with the market is usually a part of the actions performed by the governance organization: here they have been shown alone in order to emphasize their crucial role in an industrial network. Their scope, indeed, is to make as strong as possible the presence of the industrial network in the markets of final products, labour, finance, etc.

The following example shows the importance of a strong synergy among social, economical and political actors at a local or regional level, for the ability of an individual enterprise to compete in a worldwide market, thus emphasizing the functionality of the ISEE component of a network under evaluation.

Canelli – Santo Stefano Belbo is a *wine industrial district* situated on the Southern Piedmont, a region of northwest Italy. Founded in 1998, it deals with the production of two local wine brands and wine treatment machineries. In recent years the district has been included in a regional territorial development project, whose purposes include the internationalization of SMEs. There were two motivations for this: the commercialization through the Internet (e-commerce) and huge institutional commercial events with foreign delegation (hosting and visiting). A lot of effort has been made to achieve a good commercial relationship and a lot has been spent to settle an excellent technical architecture. Now the goal is to understand the results brought on by the process, the impacts it had, and how it was pursued by the district, which involves the district committee.

The Canelli – Santo Stefano Belbo district received regional funds for developing the submitted projects, because they perfectly matched the prescription of the Regional Law no. 24/97, which is "to support the local products marketing of districts products, in particular regarding to foreign markets, putting attention on trend analysis, international contracts laws, customer assistance, and looking for commercial and technological partnerships".

The emerging and common aim between all the districts' financed projects is the internationalization of their products. The commercialization in foreign markets is a goal that requires an incisive marketing action, which SMEs for economical reasons could not effort. In 2003 the Piedmont Region designated 61.2% of funds to support such types of initiatives.

In the last few years the Canelli – Santo Stefano Belbo district submitted different projects for subsidy. One project involves the search of new markets, the production technology innovation and the creation of a virtual "window", that is a website named "Vinotel". A second project concerns the coalition between enterprises to achieve the ISO 9000 Certification. Another, named "Wine and Oenology", concerns the creation of another website to inform partner firms about fiscal and financial assistance, quality certification criteria, and guide lines to cellars' environmental managing, according to the new national regulation.

Referring to the product delivery, the wine industrial district of Canelli is one of the world leaders in the packaging sector, being able to design and manufacture systems capable of accompanying any product (not just liquids) from the packing phase to handling in the warehouse and shipment. About 80% of the sales of the oenological machinery sector are sent abroad. This internationalization process has concurred in developing the internal enterprises but has not always succeeded in transmitting an image of the territorial system: it is harder to link the image of a territory to a high-tech sector; the work, experience, knowledge and qualities of which are often reserved only to the sector's operators, inasmuch as it is not always easy to present them to the general public. Therefore, institutionally speaking, the partners of the Canelli district have recently decided to create a commercial event, the Canellitaly project (see the address: http://www.canellitaly.it), to promote effective communication and opportunities for meetings which are able to link the oenologic machinery sector to the territory, putting the institutions in a position of "go-between", to bridge the gap between enterprise and new development opportunities.

Owing to the significant returns on investment made in e-commerce, the district committee is working with all partners to a new triennial project plan, to set up a consortium with public and private bodies funds. Among others, a real motivation for the consortium constitution is the necessity for improving the road and railway networks, such as to link the Canelli area with the high-level European network of road/rail channels.

The above considerations, all referred to the evolution of the component *interactions with the socio-economic environment* (ISEE) of the Canelli district, show that since 2003 there has been an improvement both in the SME propulsion to enter the district as new partners, and in the growth of the district culture and interest in the internationalization process, which is the main goal of SMEs. The projects have realized commercial benchmarks, acquisition and consolidation of new markets and marketing events. The big result the district achieved is the awareness that the internationalization process does not coincide only with exports, but also requires deep changes in how partnerships are created, in quality assurance, in technology innovation, in management class training, in business analysis instruments, and in customer relationship management.

Another example (whose data can be found on the CODESNET website: http://www.codesnet.polito.it) demonstrates the importance of the *organizational arrangement* (OA) in assuring and maintaining efficient and effective connections among the network partners.

Etna Valley is a high-tech industrial district, situated on eastern Sicily, near the volcano Etna, in the area of Catania. This industrial district includes a big multinational company (ST-Microelectronics), about 200 local SMEs, and 23 other centres from multi-national enterprises such as Nokia, Canon, IBM, Alcatel, Vodafone, etc. Local high-tech SMEs are closely connected to STM, which is a "reference point" for their activities. Presently, a centralized organization – as a reference committee or syndicate – supporting the district cohesion and development, is still lacking. However, a committee is going to be instituted in the future, with the following targets: linkage between big companies (ST and other multinationals) and local SME; exchange of view and information among the companies of the district, and more cooperation; support for the SME in financial difficulties; promotion of the district image; institution of a common service for financial and legal assistance, in order to reduce consulting service expenses of companies; match-making service for buyers and vendors; vocational training activities; management of national/community funding. The committee's political functions and the individual companies decision-making authority is in the definition phase. In Fig. 4.2 a simplified representation of a possible organization of the future committee is depicted.

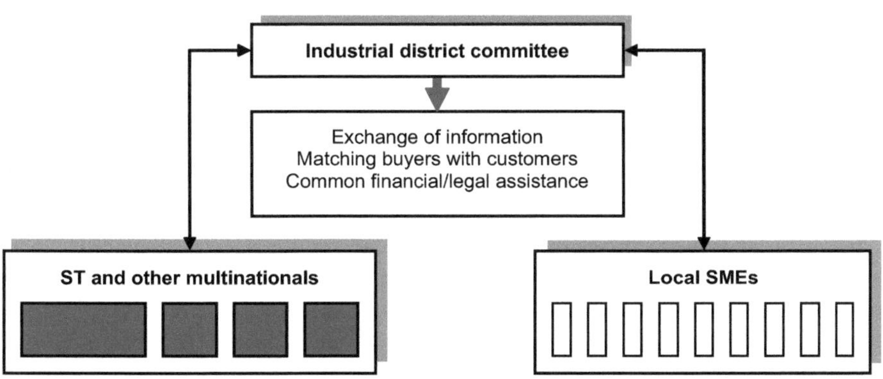

Fig. 4.2 Organization of the district committee

Both examples, clearly referred to typical types of Italian industrial districts, such to present the main characteristics of several SMEs clusters in different European countries, make evidence of the usefulness of the above-introduced "meta-model" of the network's three basic components. The previous examples only show that the meta-model allows for the "description" of the network's main functions: the following sections will address the utilization of the meta-model for "understanding" how the network functionality depends on the main characteristics of the network itself, such as to allow the analyst to evaluate the network performance.

4.2 A SME Networks Model for Performance Analysis Through Comparison

D. Antonelli and A. Villa – Politecnico di Torino

The SME networks' meta-model introduced in the previous section must be translated into a "formal model" (expressed in either logical or mathematical terms) in order to provide three components to be used by analysts – the operation structure (OS), the organizational arrangement (OA), and the interactions with the socio-economic environment (ISEE) – as specific viewpoints of the performance analysis.

The motivation of such a translation into formalism can be clearly shown by the following consideration: the possibility of individually analysing each component of a SME network will allow one to evaluate proper measures of the network performance in terms of efficiency, effectiveness and economic convenience. The network performance indeed depends as a whole on all the network's three components, because an efficient network structure, an effective governance and a good ability to interact with its external context surely are necessary conditions for a good performance. But any industrial analyst needs to be more precise: it looks incorrect to refer to the network structure to evaluate the good efficiency and effectiveness of the industries themselves.

The structure of a SME network consists of links, each one with the proper capacity for moving and storing parts, and of nodes, each one corresponding to an enterprise with proper production capacity. In other words, a structure can be characterized by proper "technical parameters", that specify both the structure type and the dimensions (as for example, number and types of links, transportation time and capacity of each one, etc.). A similar consideration could be done for the other components.

In a similar way, to measure the performance requires access to specific "performance indicators", which must be measurable, and obviously different if related either to the structure, or to the governance organization, or to the interactions with external context. And the identification and selection of these *technical parameters* (TP) as well as of the *performance indicators* (PF) usually can be obtained through a gradual and iterated approximation.

First of all, for each one of the three components of the meta-model, the "main analysis issues", the most crucial and typical questions to which any industrial analyser wants answer, must be recognized. In practice, what we call main analysis issues can be identified as follows: (a) they are questions related to either a design or a management problem concerning the network organization or development; (b) then, to answer each question, one must know on which TPs the designer/manager could act through a TP modification, and by which PIs the same designer/manager could verify the result of his TP modifying decision.

Once the main analysis issue of interest is selected, the analyst must be able to associate to such an issue some TPs (as input information about the network under analysis, for the considered issue) and at least one PI, which can measure the performance of the network. Then, considering each PI, the analyst tries to express the dependence of that PI from the selected TPs.

In this way, the analyst can formulate a correlation model to give an interpretation of the network performance, such as how one or more measures of performance depend on some structural parameters of the network is shown.

Indeed, the meta-model together with the formal models of the three component elements offers some ideas for developing an industrial user-oriented *methodology for evaluating the performance of a SME network.*

Two steps can be easily identified in such an "industry-oriented methodology".

1. Decompose the performance evaluation problem into three parts, one for each component element of the meta-model, and then of the SME network.
2. For each component element (OS, OA, ISEE), the following actions have to be made: (a) associate to the formal models of the element itself a set of measures, either referred to TPs or to PIs; (b) state clear relations between the TPs and the PIs, motivated by the formal models of the element itself; (c) associate to the component element a set of data collected in a real SME network, to be used as a benchmark.

A first approach to use the meta-model consists of representing, independently for each component element, a simple association existing between TPs and at least one PI, by considering the latter as a consequence of the former. In practice, this representation corresponds to describing "expected correlations" between the set of TPs and each PI.

In a theoretical approach, any relations of this type must be explained by formal models related to the considered component element, as mathematical models and optimization procedures.

In a practice-oriented approach, these relations could also be estimated by an industrial expert; in this case, the table of relations could be viewed as an *industry-oriented model*, meaning a correlation matrix linking TPs to PIs. A sketch of the industry-oriented model is illustrated in Fig. 4.3. Note that the definition of industry-oriented model is adopted only to evince one of its potential uses, which is to

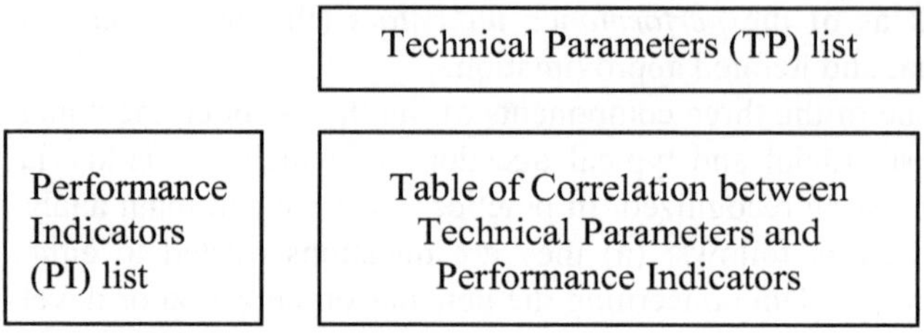

Fig. 4.3 Industry-oriented model

collect information from an expert, and then make comparisons among different networks. But this is not an approach which allows one to "understand": rather, it is utilized only for "comparing".

To illustrate a potential utilization of the above sketched correspondence between TPs and PIs, it should be necessary to have at one's disposal measured values for both parameters and indicators: some real applications will be widely presented in the following. Here some simplified considerations, based on a few realistic data available in the CODESNET Web Portal as well as in the literature, are given for the sake of understanding the industry-oriented model and its utilization.

Let us refer to a SME cluster characterized by the data of Table 4.1.

As the reader can see, some data have not been declared. Then, some PIs cannot be considered for a complete performance assessment. On the other hand, no information has been obtained for some TPs, which cannot be used either. However, the above two sets of variables (PI) and parameters (TP) can be put into relation together, by using the suggestion of an *expert*. That is, an industry-oriented model could be filled by expressing the strength of the relation between

Table 4.1 Data on a SME cluster

Performance indicator (PI)	Value
(1) Production volume	140 Parts/month
(2) Lead time	30 Days
(3) WIP; internal	Negligible
(4) Flow time	Not declared
(5) WIP; raw materials	Not declared
(6) Plant utilization	75%
Technical parameters (TP)	**Value**
(A) Demand for products	1,500 Parts/year
(B) Layout type (graph)	Six-stage line
(C) Operation times	Not declared
(D) Production capacity	180 Parts/month
(E) Material utilization	Equal to prod. capac.
(F) Cost per operation	Not declared
(G) Monitoring type	Not declared
No. Employees	About 400 pers.

Table 4.2 Linkage table

	A	B	C	D	E	F	G	Weights
1	9		5	9				30
2		5	9	5				20
3		9			5	5		10
4		9		5			5	10
5					5	9	1	10
6	9			5				20

each TP and each PI. To this aim, the expert's opinion can be translated in terms of a "vote", measuring the strength of the TP versus PI relation over a decimal scale. An example is given in the Table 4.2, where also a classification of importance of the PIs is expressed by a vector of weights.

It is easy to understand how the expert contribution could provide significant help in describing the SME cluster under examination.

In practice, each expertise-based correlation (*i*,*j*) is a grade showing how much the performance indicator PI(*i*) should be conditioned by the technical parameter TP(*j*), obviously according to the opinion of the involved expert panel. More accurately, the correlations are here stated in terms of three grades, where value 9 corresponds to a "strong correlation", value 5 "mid correlation" and value 1 "low correlation". Then, by reading the correlations in Table 4.2, the analyst can say that the PI *production volume* is *strongly dependent* on the TPs *demand for products* and *production capacity*, and it should have a *mid correlation* with the TP *operation times*, even if this last data is not available. Similar considerations could be done for the PI(2) and the PI(6), whilst no conclusion can be drawn for PI(3), PI(4) and PI(5). In summary, the industry-oriented model gives a qualitative but really useful description of the system, by giving a sketch of its structure.

A comparison between the SME network under evaluation and some other benchmarks, as for example the same industrial sector to which the network itself belongs, can be done by comparing:

1. *The two correlation matrices*, therefore to see if the network is characterized by a "structure" (i.e. relations between the performance indicators and the technical parameters) similar to the average structure of the other networked systems belonging to the same industrial sector
2. *The respective values of the performance indicators* of the network under evaluation and those of its industrial sector, in order to estimate if the network under evaluation is a really efficient-effective-convenient industrial body
3. *The respective values of the technical parameters*, such that one can estimate if the network under evaluation is a really advanced industrial body

A further improvement of the above comparison could be obtained by also using a list of weights, showing the respective importance of the considered performance indicators, according to the system manager opinion. If such a list of weights is available, a corresponding list of weights for the technical parameters can also be computed, such as to classify also the parameters' importance in describing the cluster to be evaluated. In such a way the analyst, besides the correlation model linking technical parameters and performance indicators, can also compute a classification of technical parameters: this makes for an easier comparison between alternative networks. Now it is important to analyse how such a formal modelling could be used to link TPs and PIs through mathematical relations. Indeed, a set of mathematical relations is the desired model if one aims not only to "compare" different networks, but also to "understand" why a network behaves better than another. A mathematical model is an explanatory model: it can show how a performance indicator "depends" on some technical parameters and then its value is a consequence of the values either assumed or attributed to such parameters.

4.3 A Formal Model for SME Networks Performance Understanding

D. Antonelli and A. Villa – Politecnico di Torino

Typically, a procedure for performance evaluation of an industrial system consists of two complementary actions: first, *to compare* the values of the performance indicators observed in the considered SME network with the ones of similar networks*; then, to understand why*, in the SME network under examination, some performance indicators (PIs) assume the observed values, depending on the detected technical parameters (TPs). Note that the second action is of prevailing importance: indeed, "evaluate" means "understand", and only after that, "compare".

The importance of the second action (i.e. *understand* by proper formal models) can be fully realized by the following discussion and interpretation of the model formulation proposed above, in view of its potential different utilization as a *model helping to understand.*

A fruitful application of the above-mentioned industry-oriented model comes from considering that any relation between a performance indicator PI and a set of technical parameters TPs corresponds to a design and/or management problem. This problem involves the SME network under examination, namely the problem of either designing or managing, whose solution can obtain the best value of the considered performance indicator (PI).

In this case, note that the search for the "best" value of a performance indicator PI means that the value of that PI must be obtained by finding which values of technical parameters can "optimize" the performance of the considered system. In this way, the improvement of the system performance can be "explained" by showing how the values of the technical parameters could affect the system operations.

Considering one of the component elements of the meta-model described in Sect. 4.1, the above statement means that the performance indicator usually is a "measure" associated to a formal model of the component element, while a set of technical parameters are the “input data” specifying how the element itself can operate, according to the formal model considered.

In other words, paying attention again to the formal models describing how a network component (that is either the network structure, or the network governance or the network interactions with the outside) behaves, to each relation between a performance indicators and a set of technical parameters one can associate a design/management problem of the SME network stated by the following elements:

1. A set of technical parameters specifying the problem’s *independent variables* (then, the inputs)
2. A proper performance indicator, specifying the problem’s *target* (then, the problem output to be optimized)
3. A problem *formulation* (e.g. in mathematical or logical terms) and possibly a solution procedure

Considering the table of TPs versus PIs (which, in the industry-oriented model contains the TP versus PI correlations), in the *i*th row corresponding to a given performance indicator PI(*i*), one can describe/formulate a particular problem PR(*i*) by marking which technical parameters TP(*i*,*k*) are present in the considered PR(*i*) problem itself. Then, the optimal value PI*(*i*) of the performance indicator PI(*i*) can then be obtained by solving the associated optimization problem PR(*i*) with respect to the admissible values of the related technical parameters TP(*i*,*k*), by applying a specific solution procedure, which could be referred to in the corresponding *i*th row in the table.

If so, the above-mentioned table becomes a *problem/procedure catalogue,* which connects a set of TPs to a PI, denoted as a “problem-oriented model”.

A typical scheme illustrating the problem-oriented model is presented in the following Fig. 4.4, referred to the component element operational structure (OS).

To give a clear description of above problem-oriented model, let us consider the following example (see Fig. 4.4). Here, two sets of TPs are considered, respectively

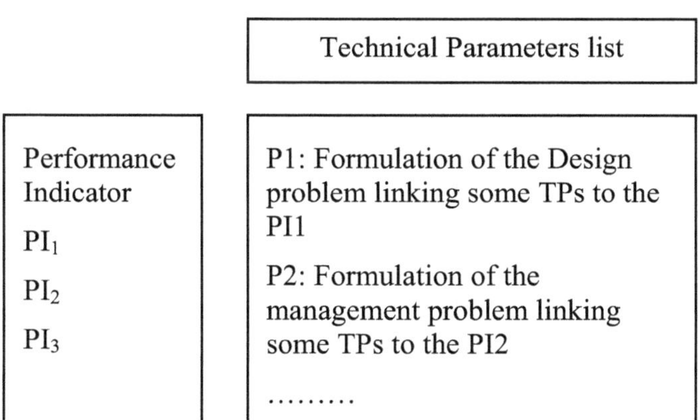

Fig. 4.4 Problem-oriented model

related to the logistics of the SME network (e.g. transport capacity parameters of the logistic links among SMEs) and to the production activities of the included SMEs (e.g. manufacturing times and production capacities of the SMEs). Two related PIs are also adopted, namely the "number of bottlenecks" in the logistic network connecting SMEs, and the "average lead time" to complete the final products.

Some examples to be considered with Fig. 4.5 are:

- PI(1) = Number of bottlenecks in the SME network
- PI(2) = Average flow time to complete final products
- TOOL(1) = Production flow analysis (PFA) procedure (Kusiak 1990)
- TOOL(2) = Longest route selection procedure (Villa 2006)

In the first case, the aim of the analyst is to evaluate if the number of bottlenecks of the SME network under examination is as low as desired. To this aim, the formal problem of optimizing the number of bottlenecks in a network of work-centres, for given external demands of final products, is considered. The optimization can be searched with respect to the loading conditions of the work-centres, depending on the production flows addressed in the network links. Then, the production flows addressing problem can be solved by applying the so-called principal flow analysis (PFA) method, based on the allocation of flows according to the matrix describing the network connections (referred to as firm-to-firm incidence matrix in Fig. 4.5).

In the second example, the goal of the analyst is to compute a reasonable value of the maximum flow time in the SME network. Then, based on a model of the whole production system as a "graph" of work-centres and the estimation of the work-centres loading conditions, a procedure to identify the longest route, starting from the input work-centre up to the output one for any given final product, can be applied.

In both cases, the figure above illustrates simple examples of the potential application of a problem-oriented model, which is a model defined such as to solve the problem (either design or management) concerning how the value of a performance indicator could be affected by some technical parameters. Obviously, it

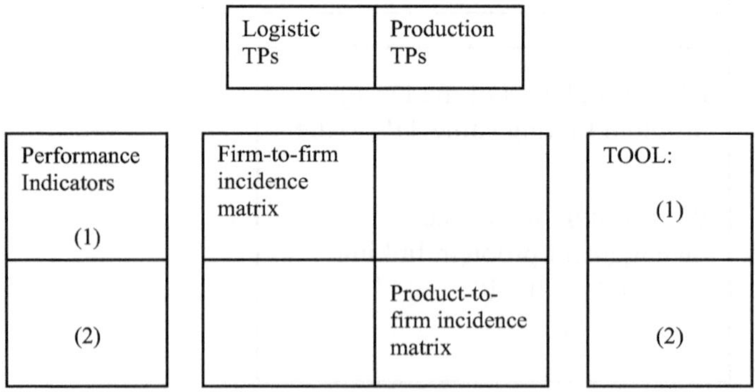

Fig. 4.5 Example of a problem-oriented model

is just a schematic example: the formulation of the two problems mentioned there implies that the two involved sets of technical parameters are known, respectively.

Besides, the introduced general formulation allows to focus some important aspects of the model and the consequent formal method for evaluating the network performance.

First, this model clearly shows that any performance evaluation implies the consideration of not only the actual behaviour of the network under examination, but also how its present operations could be improved. Indeed, the real scope of any performance evaluation is simply to find causes of an insufficient performance, if any, and suggestions for its improvement. This second action must be driven by a clear description of how to innovate the network technical parameters. And, this decision can only be adopted if a model of the network and of the innovation problem is easily available.

Second, the formulation of the problem-oriented model could appear extremely general, because it is just an expression of the potential links between a given performance indicator and some technical parameters, through a PI-optimization problem. However, in the technical literature there exists a lot of either similar or slightly different formulations of the same design/management problem and, for each formulation, a lot of solution procedures/methods. Then, if a formal methodology for evaluating the network performance is desired, *it is absolutely necessary to select a specific catalogue of the typical problems that will be considered.* This specifies a range of design/management problems by which the network performance can be analysed, and, mainly, the dependences of the performance indicators' values on the network technical parameters can be studied and justified. This clarifies that the utilization of the problem-oriented model as a performance evaluation and innovation tool requires the theoretical knowledge concerning optimization methods, as well as mathematical knowledge concerning the problem formulation and solution.

Now, the question is: is theoretical knowledge a necessary support for either innovating or managing a SME network? The real answer is yes. But the practical answer is that an industrial analyst needs to have sufficient suggestions from theoretical knowledge and methods without having the necessity of solving either too complex or time-consuming mathematical problems. This is the practical motivation of the conclusion already stated: a catalogue of the typical problems concerning either management or organization, of a SME network must be made available to the analyst, such to have at his/her disposal methods and procedures to be directly applied.

4.4 An Example of Cluster Analysis

B. Caroleo and T. Taurino – Politecnico di Torino

The two models above introduced seem to be effective tools for a performance evaluation of real industrial bodies, either enterprises or SME networks. They *seem to be*, but a practical utilization of them must be proved. So, a realistic example is introduced here, to allow the reader to understand the strength and weakness of the two models: depending on their complexity, the reader should realize how the concepts underlying such models could be possibly translated into an industrial – and then, simple and practice-oriented – analysis procedure.

The example proposed in the following is based on a specific selection of the technical parameters and the performance indicators to be associated to each network component, that are the structure, the governance and the interactions with the external.

In order to have some criteria for selecting some technical parameters and some performance indicators for any SME network to be analysed, to each network component a small set of specific questions are associated: each question identifies a performance analysis problem, that is, a problem of either network design or organization, and from the solution of which the network performance can be greatly affected.

So, in case of the *operational structure* of the SME network, the following main questions are considered:

1. How are production operations and produced volumes distributed among the network enterprises?
2. What are the different skills employed in the SMEs belonging to the network?
3. Which type and structure are the adopted logistic connections?

In case of the *organizational arrangement* (i.e. the governance) of the SME network, the three main questions are the following:

1. How are management responsibilities attributed to each enterprise?
2. How are internal agreements or control procedures negotiated among the SMEs?
3. What is the type of organization chart selected, to assure the best efficiency and effectiveness to the whole SME network?

And finally, the set of main questions associated with the third component, that is the *interactions with the socio-economic environment*, are as follows:

1. How are commercial agreements negotiated between the SME network and the external markets?
2. How is a product innovation program decided by the whole SME network?
3. Which dynamic evolution of the SME network can be forecast?

These main questions indeed are the typical problems that usually drive a performance evaluation in any industrial network. Then, they clearly define the *main analysis issues* above introduced.

To each main analysis issue, some technical parameters as well as at least one performance indicator are related, thus generating a table of correspondences between each main analysis issue, on one hand, and a set of specific technical parameters as well as a set of proper performance indicators, on the other.

The motivation can be shown by the following points:

1. Each main analysis issue specifies both a viewpoint and a problem, to which an industrial user could approach his/her own evaluation task.
2. The related set of technical parameters defines the set of information that the analyst can use to describe the network component, and then to select a component model to support any evaluation consideration.
3. The related set of performance indicators, in turn, states the measures, to which the performance of the considered component of the SME network under examination, can be measured.

The following Table 4.3 show the technical parameters and performance indicators frequently adopted in performance evaluation tasks.

Sets of technical parameters and related performance indicators suggest that a description of each network component can be done by showing the most evident *correlations between technical parameters and performance indicators*. Any correlation, which could be estimated on the basis of the experience, indicates if and how a technical parameter can affect or modify the value of a performance indicator.

The three models related to each network component (operational structure (OS), organization arrangement (OA) and interactions with the socio-economic environment (ISEE)) can be filled as in Table 4.4, where some examples of correlation value are proposed. These three correlation models can help in explaining how internal SMEs interact together and how the whole set of interconnected SMEs can be managed.

The typical application of the proposed model is to operate as a "comparison tool" for evaluating some SME networks by comparing their respective values not only of the performance indicators (the most important data), but also of their respective technical parameters and correlations tables.

As proposed in Sect. 4.2, accounting for the network component operation structure, a comparison between a SME network under evaluation and other similar SME networks can be done by comparing:

- *The correlation matrices*, such to verify if the SME networks to be compared are characterized by similar structures.
- *The respective values of the technical parameters*, such to verify if the SME network under consideration presents a structure with dimensions similar to those of the other networks (note that networks with really different dimensions, e.g. number of SMEs included, production capacities, transportation and logistics types, etc., cannot be significantly compared).
- *The respective values of the performance indicators*, such to estimate if the SME network under evaluation is an efficient (effective, convenient) industrial network.

Typically, the performance evaluation that can be drawn through comparison of a given network with either similar industrial bodies is crucially based on the analyst's experience: the correlation values can only be chosen by experienced persons, having a clear view of the benchmarks to which the system under examination could

Table 4.3 Examples of technical parameters and performance indicators

Operational structure (OS)		
Main analysis issues	Technical parameters	Performance indicators
How are production operations and volumes distributed among the enterprises?	1. Spatial allocation/ distribution of partner SMEs in a geographical area	A. Degree of concentration[1] of production
	2. Max production capacity of each partner SME	
What are the different skills employed in the different enterprises?	3. Division of labour[2] between partners	B. Personnel employed per partner
What type/structure of logistic network is used?	4. Organization of internal logistics = Type of network of physical flows[3]	C. Number of logistic bodies/transport Means
Organization arrangement (OA)		
Main analysis issues	Technical parameters	Performance indicators
How are management responsibilities attributed to each enterprise?	1. Responsibility[4] assigned to each partner SME	A. Type of organization structure[5]
How are internal agreement or control negotiated?	2. Internal agreements and control mechanisms	B. Type of coordination strategy/function
	3. Types of collective agreements with external firms	
What type of organization chart is selected, to assure the best efficiency and effectiveness?	4. Information pattern[6] at disposal	C. Type of coordination body[7] (if any)
Interactions with the socio-economic environment (ISEE)		
Main analysis issues	Technical parameters	Performance indicators
How are commercial agreements negotiated?	1. Types of commercial agreements with clients and suppliers	A. % coverage of the product market
How is a product innovation program decided by the enterprises?	2. Types of production strategies[8]	B. % employees over population resident in the area
Which dynamic evolution of the network can be forecast?	3. Types of innovation plans	C. Investments in innovation plans

[1] The degree of concentration of production corresponds to the percentage of local production controlled either by the whole networked enterprise or by its first most important SMEs.

[2] Each SME could be oriented to specialize its production process by just one phase or a few phases. The SMEs belonging to a MBN-Enterprise usually operate in the same industrial branch,

but they could be "upstream firms", "downstream firms" or "ancillary firms", depending on their specialization.

[3] "Type of flows network" is a description of the physical flows among the SMEs, in terms of graph type, SMEs storage capacity, and links' transportation capacity.

[4] "Responsibility" refers to the type of management decisions that the partner SME can take, and the set of other partner SMEs, which depends on those decisions.

[5] The "type of organization structure" has to be chosen among the existing few number of structures usually applied in industrial bodies, i.e. start-structured or hub-and-spoke type, etc.

[6] The "information pattern" is the description of the information at the disposal of a SME, concerning the state of other SMEs and of the links connecting itself to the others.

[7] Typical example of coordination bodies could be a committee, a managerial centre, either dedicated to a political coordination or managerial coordination.

[8] As examples: MRP-based or JIT or OPT.

Table 4.4 Example of the correlation-based industry-oriented model of a SME network

Technical parameters Performance indicators	1. Spatial allocation	2. Labour division	3. Logistic organization	4. Prod. capacity of each partner
A. Degree of production concentration	MID	HIGH	MID	HIGH
B. Personnel employed/ partner SME		HIGH		MID
C. Number of logistic bodies/transport means	MID	MID	HIGH	

Technical parameters Performance indicators	1. Assigned partner responsibility	2. Information pattern	3. Internal agreements	4. External agreements
A. Type of organization structure	HIGH	MID	HIGH	
B. Type of coordination body			HIGH	MID
C. Type of coordination strategy	HIGH	HIGH		MID

Technical parameters Performance indicators	1. Commercial agreements	2. Production strategies	3. Labour cost and labour market constraints
A. Coverage of the product market	HIGH	MID	
B. Employees over population in the area		MID	HIGH

be compared. And a correct filling of the correlation matrix is surely the crucial step in the performance evaluation. So, in the case of a young analyst, this step could become an insuperable barrier.

On the other hand, even if done by experienced people, a performance only based on comparison could be insufficient to give a transparent explanation of the real causes of weakness – as well as of strength – of a SME network. These industrial bodies, indeed, are so complex and their management so difficult, that to understand potential motivations for an incoming crisis is an extremely difficult task.

To receive more suggestions for better understanding the dynamics of a SME network, an analyst needs to have available a formal model of the three components introduced above: this idea has been the motivation of the previous Sect. 4.3.

If one takes into account the existing correlations as an "expression of dependence" of a performance indicator from some functional parameters, then considering the row of the table referred to a given performance indicator could suggest that some parameters' selection problem could be stated to identify the desired correlations.

But it must be noted that "parameters' selection problem" means, in a theoretical viewpoint, *problem of optimizing some indicator with respect to the values that could be assumed by the technical parameters*. In other words, to have significant information about the relations between a given performance indicator and a set of technical parameters one has to state a formal problem of maximizing (or minimizing) the performance indicator, when the technical parameters are the unknown variables of the optimization.

As a consequence, an evaluation driven by a formal model of a network component will be based on the adoption of the mathematical statement of a problem of optimizing a given performance index, with respect to a given set of technical parameters, and for a given model of the network component. This last one will describe how the technical parameters could affect the performance indicators value, thus playing the role of "constraints" in the optimization problem.

In this line, the industry-oriented model, even if only containing qualitative correlations as the one described above, has the scope of indicating which are the most significant links between a performance indicator and some technical parameters. For each one of these significant relations, the problem-oriented model must specify the optimization problem to be approached. This is the case of the following table, containing the optimization problems corresponding to the suggestions of the above industry-oriented model.

By summarizing the above considerations, the *milestones* of a complete methodology for SME network performance evaluation can be evidenced as follows.

The table containing the typical *main analysis issues* and, for each one of these, a set of TPs and at least one PI, will be the "driver" of the analyst in selecting the optimization problem to be approached and the required information and data. In particular, the correlation matrix of an industry-oriented model will supply the analyst with information (often only qualitative) on how each TP could affect a given PI, and will specify which real data have to be measured in the network in order to obtain a per-

Table 4.5 Example of the problem-oriented model of SME network corresponding to the industry-oriented model of Table. 4.4

Technical parameters Performance indicators	1. Spatial allocation	2. Labour division	3. Logistic organization	4. Prod. capacity of each partner
A. Concentration of production	Select/optimize the degree of concentration of production, with respect to the spatial allocation of partner SMEs, the division of labour among partner SMEs, the organization of internal logistics, the production capacity of each partner SME			
B. Personnel employed/ partner		Select/optimize the personnel employed per partner SME, with respect to the division of labour	→	*and* the production capacity of each partner SME
C. Number of logistic bodies/ transport means	Select/optimize the number of logistic bodies/transport means, with respect to the spatial allocation, the division of labour among partner SMEs, and the organization of internal logistics			

Technical parameters Performance indicators	1. Assigned partner responsibility	2. Information pattern	3. Internal agreements	4. External agreements
A. Type of organization structure	Select the type of organization structure, with respect to the responsibility assigned to each partner SME, the information pattern available, and the types of internal agreements and control mechanisms			
B. Type of coordination body (if any)			Select the type of coordination body, with respect to the types of internal agreements and control mechanisms, and the types of collective agreements with external firms	
C. Type of coordination strategy or function	Select the type of coordination strategy/ function, with respect to the responsibility assigned to each partner SME, the information pattern available			*and* the type of agreements with external firms

Technical parameters Performance indicators	1. Commercial agreements	2. Production strategies	3. Labour cost and labour market constraints
A. Coverage of the product market	Select the best production program, with respect to the types of commercial agreements with clients and suppliers, and the types of production strategies		
B. Employees over population in the area		Select the best percentage of employees over population resident in the area, with respect to the types of production strategies (that could be applied), and the labour cost and labour market constraints	

formance estimation. On the other hand, the matrix of the related problem-oriented model will indicate to the analyst how he/she could search for answers to the questions contained in the main analysis issues (Table 4.5). Each optimization problem, indeed, is referred to a specific main analysis issue and, by searching for its solution, one can derive suggestions on either a procedure or some criteria for correctly analysing the SME network and estimating correlations among TPs and PIs.

These considerations can be summarized in an outline of the performance analysis methodology.

4.4.1 Basic Concepts for SME Network Performance Analysis

Independently for each component element of a SME Network, namely operation structure, organization arrangement, interactions with socio-economic environment, the following can be said:

1. The related problem-oriented model describes which formal models can be used to describe the dependence of a performance indicator (PI) from a set of technical parameters (TP), and indicates which measures must be done.
2. The corresponding industry-oriented model, to be filled eventually with the help of an industrial expert, can be used to validate the interactions between the considered PI and the set of TPs explained above, in order to justify the adopted formal models and associated computation procedures.
3. The set of pairs <TP,PI>, whose values have been estimated as in point 2 and whose interactions have been formally justified as in point 1, can be used as the list of attributes of the SME network, to be considered for comparison with other either similar or best-practice networks.

4.5 The SME Network Analysis Methodology

D. Antonelli, B. Caroleo and T. Taurino – Politecnico di Torino

The design of a performance evaluation methodology specifically oriented to industrial end users has been the main goal of the CODESNET project. The main idea on which this methodology has been based is the concepts of "main analysis issues", introduced in previous sections. As mentioned above, for each model of a SME network component (that is either the operational structure, the organization arrangement or "governance" and the interactions with the socio-economic environment), three typical problems have been identified, and each one stated in terms of a question of practical industrial interest.

To each main analysis issue, proper sets of technical parameters and performance indicators have been referred, such that an industrial user could try to give an answer to the "question of industrial interest" concerning the considered analysis issue, by using values of such performance indicators and related technical parameters.

In practical terms the performance of a SME network will be evaluated according to the following three steps:

1. To account for at least one PI and a related set of TPs, for each component of the SME network meta-model
2. To evaluate how each considered PI is correlated to the set of related TPs, by estimating how strong or weak is the correlation
3. To understand the dependence of each PI from the set of related TPs, in formal terms

From such steps, two consequences can be obtained:

- From the first two steps, a comparison between different SME networks, one under evaluation and some others, used as reference terms, can be done such as to know which network is the most profitable.
- From the first and the third steps, a justification of the performance of the considered SME network can be derived, as well as of the difference between the considered and the reference ones.

These considerations suggest how a methodology of practical interest should be developed, based on two main concepts:

1. The evaluation of each SME network component must be performed on a few but really significant performance indicators, and has to be justified; which means that the analyst must be able to explain the values of the most significant performance indicators based on formal models.
2. Any evaluation of the performance must be done by having available simple theoretical supports that should drive the analyst in explaining the dependence of each performance indicator from the technical parameters.

The logical consequence of the basic concepts is the following:

- From the first concept it follows that only a few key performance indicators (KPI) should be taken into account, where KPI is a measurable index that could be either estimated by real data or optimized by solving a given optimization problem.
- From the second concept it follows that for each KPI the analyst should have at his/her disposal a procedure to solve the optimization problem, such as to give a simple description of the dependence of the KPI value from those of the technical parameters.

With reference to the first point, it looks necessary to use a short list of KPIs and TPs related together.

Then, a first type of suggestion could be obtained by the analyst from a table of correspondences among the following:

- Each main analysis issue, for each network model component
- A concise description of the principal component parts of the SME network, thus modelling the components' operations
- A set of models and technical parameters of the component itself

This first suggestion can be practically obtained from Table 4.6.

The types of suggestions can be read from the table as follows. To each SME network component, for each related main analysis issue, the analyst can find some indications to describe the SME network, and then to formally model the network components. As an example, the main analysis issues are related to the SME network OS (the first set), indications to describe the network type and the internal logistics are associated (second column), and the models' indications for both are reported (third column). In practical terms, the operation structure of the SME network could be modelled by a graph where production flows model the internal logistics.

Considering the OA, the existence of a governance committee or of a leading firm in the SME network appears to be the principal characteristics, so that either the organizational chart or the leading activity of the main firm must be modelled.

The third component, the ISEE, is related to issues concerning agreements with the product market, but also suggestions to analyse both the personnel skills and the network innovation programs.

This table, indeed, presents some supports for the analyst, but is mainly related to the description of the SME network and its parts. A second type of suggestion could be surmised by considering a second version of the table, including some PIs, namely the most significant for any evaluation, i.e. the key performance indicators (KPI).

The motivations for this *second version of the table only including KPIs*, are quite evident. KPIs are quantifiable measurements, agreed to beforehand, that reflect the critical success factors of an organization. They will differ depending on the organization: a business unit may have as one of its KPIs the percentage of its income that comes from return customers; a KPI for a social service organization might be the number of clients assisted during the year. Whatever KPIs are selected, they must reflect the organization's goals, they must be keys to its success, and they must be quantifiable (measurable). In addition, the set of KPIs must be fully self-explainable: then, this set cannot be too large, and each KPI inside must be "statistically independent" from the others.

KPIs must reflect the organization's goals; therefore, an organization that has as one of its goals "to be the most profitable company in our industry" should have KPIs that measure profit and related fiscal measures, whilst an education institution, not concerned with making a profit, must have different KPIs.

KPIs must be quantifiable: if a KPI is going to be of any value, there must be a way to accurately define and measure it. It is also important to define the KPIs

Table 4.6 Examples of main analysis issues for each network component

Main analysis issues	Correspond to the following main "principal parts" of a SME network	Which can be described by the following models and technical parameters (TPs)
Operation structure		
How are production operations and volumes distributed among the enterprises What are the different skills employed in the different enterprises Which logistic network is used	Type: supply chain, production network, service network, scientific park, SME association, development agency Logistics (internal service or outsourcing)	SMEs network, with TPs: SME production rates, SME costs and production times Network of flows and queues structures, with TPs: transportation capacities, storage locations and places
Organizational arrangement		
How are management responsibilities attributed to each enterprise and how information is transferred and managed How are internal agreements, or control mechanisms, negotiated Which organization chart or coordination strategy is selected to assure best efficiency/effectiveness	Existence of leading firm(s); Governance committee: type (political, support, management); interactions with local political bodies and/or enterprise associations	Leader enterprise, with TPs: target of profit/costs, Organizational chart, with TPs: constraints among SMEs Coordination vs. competition: incentives and fines
Interactions with socio-economic environment		
How are commercial agreements with external bodies negotiated for max profit for the network How is a network innovation program decided by partners (and negotiated with financiers) Which dynamic evolution of the network can be forecasted	Agreements with external; Personnel skill levels; and qualification; Innovation programs through interactions with universities and RTD centres	Rules for external agreements, with TPs: Incentives to commonalities; Skill competence profile, with TPs: competence costs, advantages, etc. Innovation plans, with TPs: Rules to promote new products and processes; Investment target

and stay with the same definition from year to year. For a KPI of "increase sales", you need to address considerations like whether to measure by units sold or by dollar value of sales.

Many things are measurable. That does not make them key to the organization's success in SME networks. In selecting KPIs, it is critical to limit to those factors that are essential to the organization reaching its goals. It is also important to keep the number of KPIs small just to keep the attention of every SME in the network focused on achieving the same KPIs. That is not to say, for instance, that the SME network will have only three or four total KPIs. Rather there will be three or four KPIs for the whole SME network and all the SMEs within it will have three, four, or five KPIs that support the overall network goals and can be "rolled up" into them.

If a KPI of the SME network is "increased customer satisfaction", that KPI will be focused on differently in different SMEs. The SMEs devoted to manufacturing and assembling final products may have a KPI of "number of units rejected by quality inspection", while the SMEs mainly devoted to sales management may have a KPI of "minutes a customer is on hold before a sales representative answers". Success by the sales and manufacturing SMEs in meeting their respective KPIs will help the whole network meet its overall KPI. According to above statements, when considering a SME network, two complementary approaches have to be adopted in defining KPIs for the whole network and for the component SMEs.

Accounting for individual SMEs, each one can be evaluated by typical KPIs for enterprises. Several performance evaluation approaches and related sets of KPIs exist in the technical literature, as for the SCORE method (see http://www.supply-chain.org) and the scorecard approach (Bullinger et al. 2002).

For what concerns the whole SME network, the choice of meaningful KPIs must be done on the basis of the network meta-model. Indeed, this is the description of the network as a whole, by specifying its own three functions: *produce* (by using the OS), *manage* (by applying its OA) and *negotiate* with external bodies (through the ISEE). Table 4.7 shows the potential links in the CODESNET Portal.

The utilization of this second version is quite evident. The second column includes, for each set of main analysis issues related to a SME network component, a set of TPs to be either measured or specified; the third column contains a few related KPIs. Then, from each row, a proper optimization problem to be associated to the SME component of the same row can be recognized.

Now, the questions posed in previous sections have to be reconsidered, in order to find an answer of real industrial interest: how does an industrial analyst approach and solve a formal optimization problem, if a complete evaluation of the SME network performance is searched for?

In the case of a positive reply, it must be noted that the complexity and the time required to solve the performance evaluation task by using the concepts and the approach proposed here become so large as to prevent any industrial effective utilization. Thus, it appears an "industrial utilization" of the above table is mandatory to allow reasonably accurate but fast evaluations.

The starting point, as from the CODESNET approach, again is the meta-model of the SME networks and its three components, together with their main analysis issues. But, to make the table an industrial tool, a typical industrial concept must be introduced: "standardization". It looks now necessary to introduce a standard format containing both TPs and PIs for describing a *catalogue* of real SME networks, to be used such as "*cavies*" (Guinea pigs) in a laboratory dedicated to comparing networks as well as to recognize main networks characteristics.

In developing the CODESNET project, each standard format for a network representation in terms of TPs and PIs (Table 4.7) has been referred to as Virtual Laboratory (V-LAB) format, and has been stated exactly according to the list of three network components, and the three main analysis issues for each network component. Each V-LAB then contains a photo of a SME network.

However, a catalogue of V-LAB formats could only allow an evaluation of the performance of a SME network by comparison of TPs and PIs values, as it has been shown in Sect. 4.4. The analyst indeed wants to "understand why" a SME network presents a detected performance, both in case of a good PI or otherwise. This need calls for the second type of information required by the analyst: with the catalogue of V-LAB, the industrial analyst wants to have at his/her disposal also an archive of technical reports, each one offering a standard description of a procedure to solve either a given management or organization problem. Then, what is required is a Virtual Library of technical reports, each one briefly summarized in a standard format, referred to as V-LIB. Each V-LIB will be classified according to the same three network components and, for each component, the set of three main analysis issues as for the V-LABs.

Based on these two sets, namely of V-LABs and V-LIBs, the potential industrial utilization of the table of support tools for network performance evaluation can now be easily explained; the effective use in the CODESNET project will be presented in Chap. 5.

When an industrial analyst approaches the evaluation of a SME network, he/she must first approach the initial task of collecting data concerning the TPs of the network under examination. In addition, if possible, he/she should also collect data concerning the PIs, at least the estimated values of the KPIs. Once this first task has been done, the analyst can have available all data to be included in the new V-LAB, describing the SME to be evaluated.

After that, the analyst has to select the network component and the related main analysis issue on which he wants to focus the analysis; usually, it should be the component where some defects or weakness of the considered SME network have to be identified, by searching for an answer to the question referred to the considered main analysis issue in order to promote a real network innovation.

The selection of the network component and main analysis issue of interest will drive the next actions of the analyst.

- On one side, he has to search for "similar networks" in the Virtual Laboratory, i.e. networks having similar structure or organization of the considered component, in order to present good characteristics in the light of the considered main analysis issue. If so, each selected network in the Virtual Laboratory could play the role of "benchmark" for the network under evaluation. Then, a first comparison of the considered network and the benchmarks just selected could be done.
- On the other side, the analyst must also search for technical reports giving him/her suggestions and procedures to understand how the considered component of the SME network is under examination; and in particular the network characteristics specified by the main analysis issue adopted for driving the search of similar

Table 4.7 Main support tools for SME network performance evaluation

Main analysis issues	Can be described by the following models TPs	Correspond to the following SME network KPIs
Operation structure		
1. How are production operations and volumes distributed among the enterprises 2. Which are the different skills employed in the different enterprises 3. Which is the logistic network used	A. SMEs network, with SME production rates, SME costs and production times B. Network of flows and queues structures, with transportation capacities, storage locations and places	A. No. SMEs B. Logistics KPIs: Internal delivery time, inter-SME flow time Lead time vs. clients
Organizational arrangement		
4. How are management responsibilities attributed to each enterprise and how information are transferred and managed 5. How are internal agreements, or control mechanisms, negotiated 6. Which is the organization chart or coordination strategy selected to assure best efficiency/effectiveness	C. Leader enterprise, withtarget of profit/costs, D. Organizational chart, withconstraints among SMEs,coordination vs. competition	Number of leading firms Type of governance committee Note that these two KPIs must be measured through suitable grades.
Interactions with socio-economic environment		
7. How are commercial agreements with external bodies negotiated for max profit for the network 8. How is a network innovation program decided by partners (and negotiated with financiers) 9. Which dynamic evolution of the network can be forecast	E. Rules for external agreements, with incentives to commonalities F. Skill competence profile, with competences costs, advantages G. Innovation plans, with rules to promote new products and processes, investment target	No. external agreements Average (or top) skill level Investments in innovation programs (rated over the annual income)

V-LABs,could be improved. This second search is done by considering which technical reports in the V-LIB are related to the main analysis issue of interest: each one of these reports will provide the analyst support (either procedure or model or method) in order to "understand" how the considered network could be improved.

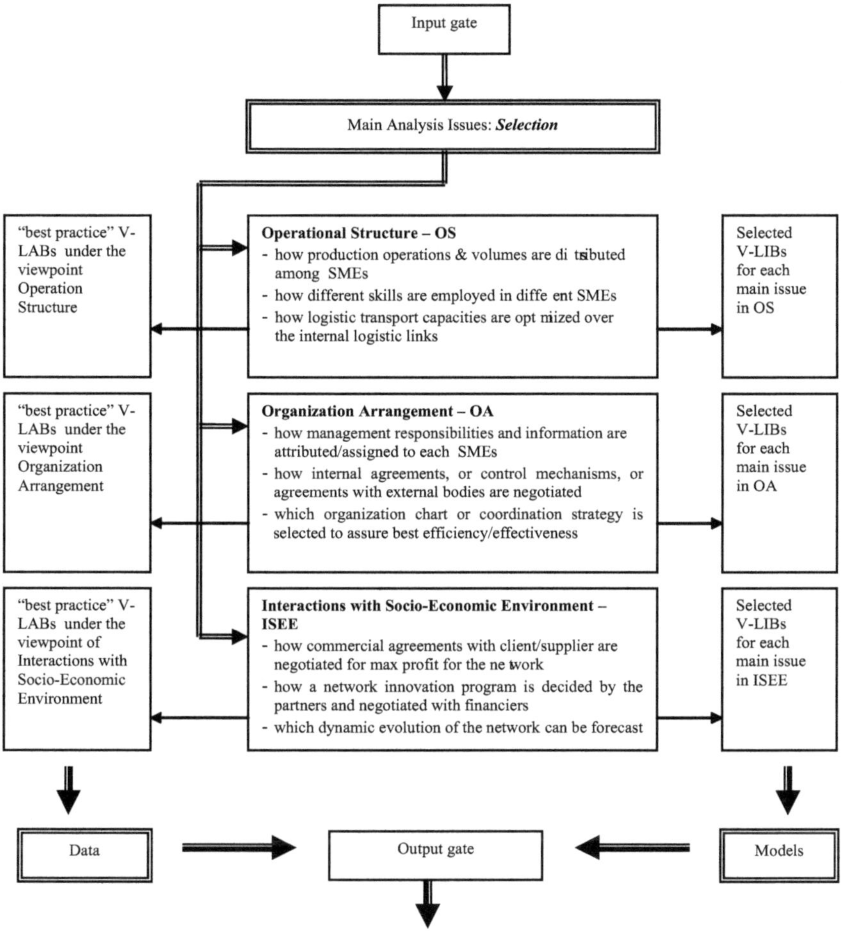

Fig. 4.6 Scheme of the proposed methodology

This reasoning methodology is illustrated in Fig. 4.6, which gives an immediate interpretation of the performance analysis methodology, according to the description above outlined.

Once the network component and the related main analysis issue have been selected, the analyst will be driven both in the Virtual Laboratory and in the Virtual Library, down a given "road" to search for similar networks, for the best possible comparisons, and also to find useful technical reports, for suggestions to understand the performance weaknesses and strengths of the considered network. Chapter 5 will show the implementation and the main results obtained.

References

Bullinger HJ, Kukner M, Van Hoof A (2002) Analyzing supply chain performance using a balanced measurement method. Int J Prod Res 40:3533–3543
Kusiak A (1990) Intelligent Manufacturing. Wiley, New York
Villa A (2006) Performance Analysis of Industrial Production Systems. CLUT, Torino

Chapter 5
The CODESNET Website: a Portal Supporting SME Network Innovation

Abstract The final goal of academics often appears to be to have at their disposal a methodology, but this was not the scope of the CODESNET project. To develop the performance evaluation methodology – presented in Chap. 4 – has only been a preliminary step. The real goal of the project was to define and implement a simple but untraditional website with the direct involvement of the two potential end users: (1) people involved with academia, to organize the website structure such as to allow simple comparisons among catalogued SME networks, and an easy search for reports which could help to understand the networks main features and (2) people involved with industry, to define both the information on SME networks to be catalogued, and the most interesting topics for selecting the reports to be included in the website. The two complementary actions of these groups of project partners gave rise to the website presented in this final chapter.

5.1 The CODESNET Website Structure

T. Taurino – Politecnico di Torino and N. Pasquino – Università di Salerno

CODESNET Web Portal (http://www.codesnet.polito.it) is a virtual place where universities and industrial bodies, interested in collaborative networks issues, can exchange information and interact. The main purpose is to provide a common supporting tool for industry and academy in selecting the suitable procedures for solving design and management problems and for evaluating the network performances. It has been developed in order to supply the end user with some performance evaluation tools, like models and methods for an effective reorganization of networks, as well as a database of case studies collected from industrial clusters and networks.

The CODESNET Web Portal is designed mainly to satisfy the industrial user demand. For this reason, in the portal construction, the main effort was devoted to making a friendly and interactive platform that allows an easy exchange of information, models and expertise between academic subjects and companies.

The main idea for the web portal construction indeed, was to create a *virtual institute* in which both theoretical and practical problems could be analysed and described. The virtual institute is organized according to the metaphor of a university institute in which the main research activities are performed in a laboratory and library; developed studies and produced works are presented in classrooms; a reception is a space for contacts together with an information desk; a coffee corner allows people to meet together for a break. This structure is reproduced in the web portal, which contains virtual rooms with the same name as those mentioned as common rooms in universities.

In the first virtual room, referred to as the Virtual Library (V-LIB), procedures and abstract models provided by academic experts are described in order to open the academic world to the enterprises. In the second virtual room, the Virtual Laboratory (V-LAB), descriptions of existing cluster and supply chain networks of enterprises can be found. These documents allow industrial people to analyse, compare and evaluate a proper industrial body (usually, a SME network).

As outlined before the CODESNET Web Portal can be used as a support for finding information, benchmarking, and models concerning supply chains and industrial networks. The purpose of this section is to give an example of the utilization of this instrument.

5.1.1 The CODESNET Web Portal

According to the paradigm of a university institution, the home page of the CODESNET Web Portal could be reviewed as the main entrance of the virtual institute. This "entrance" page, however, has been designed to easily address not only an academic end user, but mainly an industrial end user. First of all, suggestions concerning "how to use" the web portal can be found, as well as information concerning the CODESNET project. Then, the user is guided towards six "gates", as shown in the opening page illustrated in Fig. 5.1.

From the home page it is possible to access the following areas, now simply listed and then better described in the following:

1. Reception. This page contains the addresses of the CODESNET partners in order to give a reference group for clarifications, questions and discussions with involved experts.
2. Library. This page contains a very friendly library of supply chain management solutions in terms of model and conceptual scheme, provided by academic experts.

Fig. 5.1 CODESNET home page

3. Laboratory. This page contains a benchmark of clusters and supply chains, each one presented with the same, friendly, formalism.
4. Classroom. This page contains various teaching-level documents useful for knowing something more about demand and supply networks issues.
5. Registration office. This page offers the format for a registration as CODESNET supporter.
6. Coffee corner. A section containing announcements of new events.

The two core sections of the CODESNET Web Portal are the Laboratory and Library in which an end user can find V-LIB and V-LAB documents. Their conceptual structure and their scope have been described above, where the purpose now is to give a guideline for their utilization.

The "gate" of major interest for an industrial end user is opening the "Laboratory".

In the Laboratory main page (see Fig. 5.2) the three main points of view used to analyse a SME network are listed and explained through their main analysis issues. There are many reasons for this list of issues. First, this gives the end user the opportunity to find in a simple way and in simple words a description of the problems he/she wants to analyse and study. It also gives a catalogue of every aspect to be taken into account for the network description and analysis. This list also organizes the large amounts of documents collected on SME networks in order to address the search of desired issues.

In the Laboratory section, every main analysis issue is linked to the sub-set of V-LABs, which describe SME networks having as a strength characteristic the same issue. Thus, by selecting an issue, the user will automatically be forwarded to the page containing a list of these documents, as in Fig. 5.3.

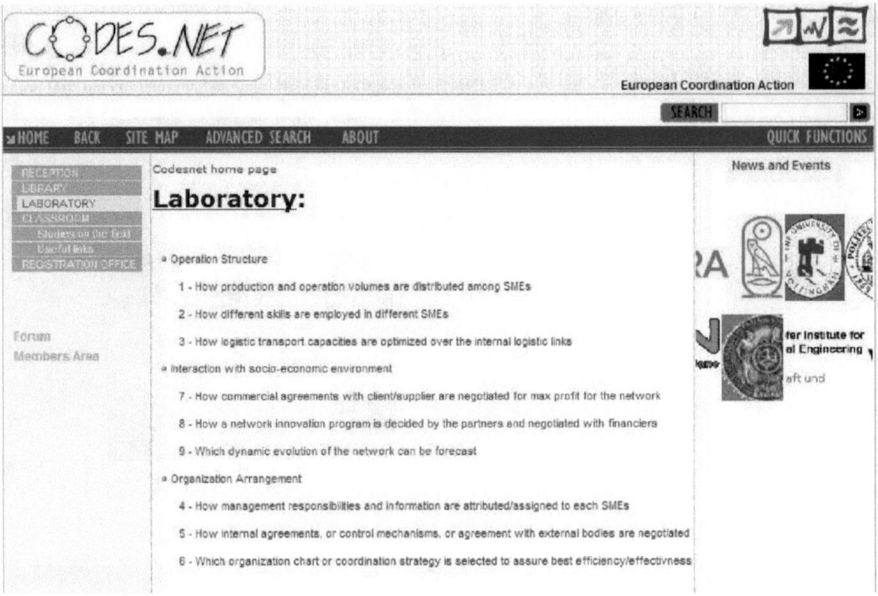

Fig. 5.2 Laboratory main page

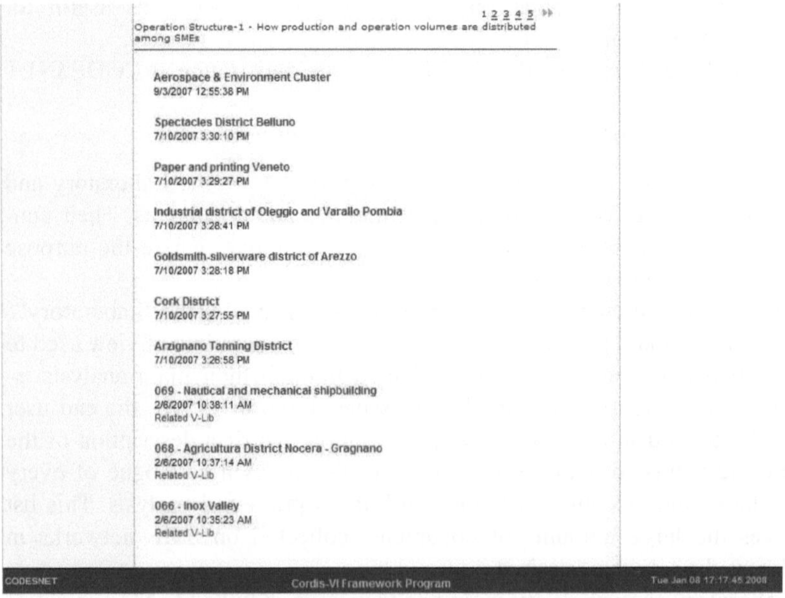

Fig. 5.3 List of V-LAB documents

The other important "gate" which can be opened is towards the Library, where the same list of main analysis issues are listed (see Fig. 5.4). Each issue of the list is linked to a sub-set of V-LIB documents. In this sub-set of V-LIB documents, as explained in Chap. 4, the end user can find a short and schematic description of scientific papers available in literature. These documents may concern surveys, case studies or general methods and models for analysis and performance evaluation: an example of the V-LIB list resulting from the choice of a main analysis issue is shown in Fig. 5.5.

The same taxonomy utilized both for V-LIB and V-LAB documents, beyond the requirement to follow the same logic and the same meta-model concepts, is essential to allow for a joint utilization of the two families of documents: to find theoretical information from the library section concerning a particular analysis issue and related examples of SME networks which could be considered as benchmarks from the same analysis issue.

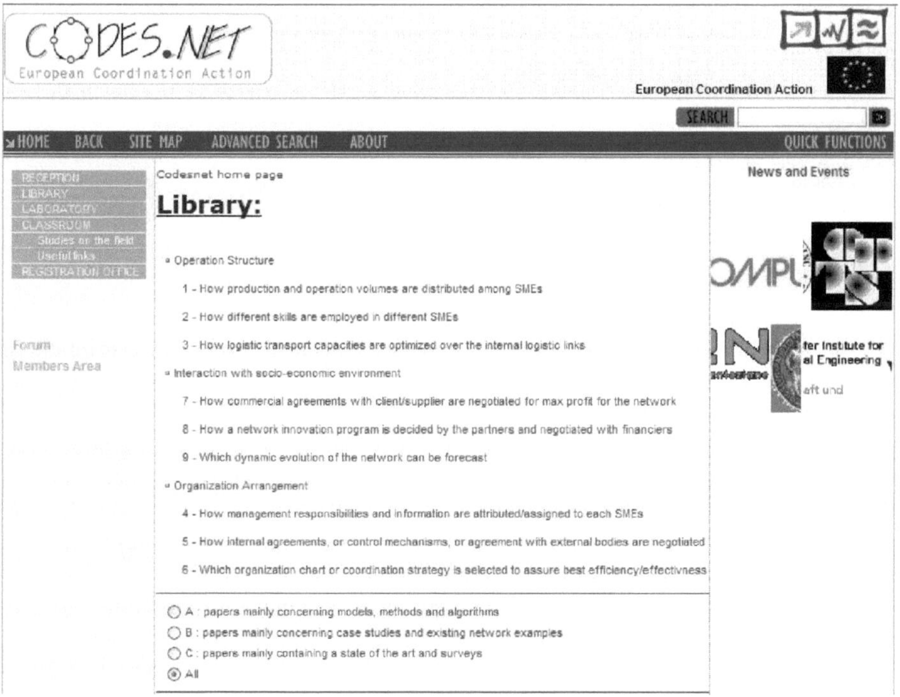

Fig. 5.4 Library main page

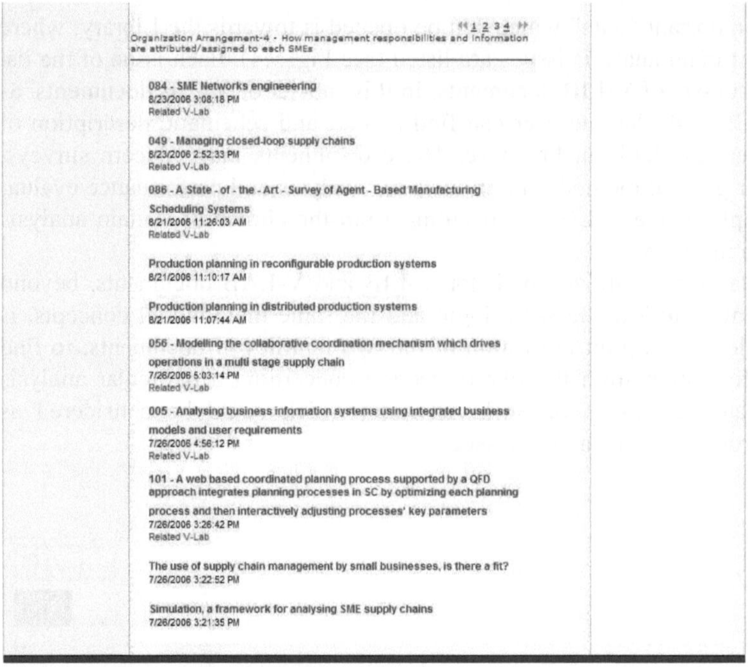

Fig. 5.5 List of V-LIB documents

5.1.2 Virtual Laboratory

As the web portal is referred to users with an industrial target, a description of what can be found in the Laboratory and an analysis of the V-LAB documents collected there is mandatory.

A V-LAB document is a sort of organized collection of information about network characteristics in order to make a complete system description according to the introduced meta-model (Chap. 4).

For this reason, in the V-LAB format, every component of the model is detailed through a list of properties and performance indicators.

The first part (Fig. 5.6) of the format contains a structure for collecting general information about the network position and contacts. For the project's purposes, it is necessary to give every user of the web portal information about how to contact a representative of the network described and presented in the document. A second but important goal is to also create a platform of knowledge dissemination about existing networks.

Then a short list of keywords and issues to identify the network activity is presented. The three aspects underlined in this list of characteristics are type of product, sector of activities and terms of process as illustrated in Fig. 5.7.

(A) DESNET (DEmand & Supply NETwork) Identification	
Name	Use Arial 14
Address	
Web site	make a link to the web site
E - mail	make a link to e-mail

Fig. 5.6 Heading space of the V-LAB format

Identification			
Keywords – issues please indicate			
English			National language (English)
product	software & application		
	information & description		
	method		
Sector	automotive		
	money & right		
	chemistry		
	advertisement & media		
	nature & environment		
	architecture		
	biology		
	electronic & computer		
	mechanical engineering		
	education & science		
	authorities & federations		
	aerospace		
Terms of process	engineering		
	manufacturing		
	Research & development		
	service		
	marketing		
	production		
	design		
	controlling		
	cooperation		
	communication		
	Coordination		

Fig. 5.7 V-LAB keywords. Issues to specify the network's industrial sector

After that, the core of information required begins with a link to the main CODESNET issues, in order to underline dimensions and features in which the industrial network appears to be particularly strong (Fig. 5.8).

A short description of the network is necessary to immediately give an overview of the kind of production and organization in order to allow for the reader to immediately evaluate his/her interest in the information contained in the following tables of the V-LAB.

The section of the V-LAB format shown in Fig. 5.9 aims to underline aspects that make it interesting from an organizational or structural point of view, by showing peculiarity and strong points of the presented industrial body.

Links to the Main CODESNET Issues			
Operation Structure	distribution of production operations & volumes		
	employment of different skills		
	optimizing of logistic transport capacities over internal links		
Organization Arrangement	communication of management responsibilities and information		
	negotiation of internal agreements/ control mechanisms, or agreements with external bodies		
	selecting of assure best efficiency/ effectiveness of organization chart or coordination strategy		
Interactions with Socio-Economic Environment	negotiation of commercial agreements with client/supplier for max profit for the network		
	deciding of a network innovation program by the partners and negotiated with financiers		
	forecasting of dynamic evolution of a network		
Short description of the industrial DESNET and of its product and/or services			

Fig. 5.8 Main CODESNET issues characterizing the network presented in the V-LAB

Links		
	English	***National language***
Strong Points explained		
Further information	Further information on the industrial body here presented, to offer a wider description of the organization and the products/services.	

Fig. 5.9 Strongest points of the network presented in the V-LAB

A "Network Overview", providing information about the network activities, is the first set of information. As shown in Fig. 5.10, general information includes network type, typology of the skill employed, percentage of market covered, while "Performance Indicators" refers to estimation of annual sales, export volume, etc.

After the network overview, each dimension of the meta-model is been detailed through three questions, in order to provide a complete as possible description of the system.

Network Overview	
General Information–Qualitative Information	
Network type	Indicate here if cluster, district, scientific park, etc...
Distribution on the region	Spatial allocation of the firms of the network in a bounded geographical area (province, municipalities, etc.)
Personnel / skills	Main skills employed in the network, their qualification levels.
Coverage of product market	Percentage of the local, national, European markets covered by the network.
Performance Indicators–Quantitative Infomation	
Estimated annual sales	
Area description	Description of the area where the network is located, number of firms, numbe of total employees in the area, number of resident people.
Average firm dimension	In Persons per firm
Export volumes (annual) of the DESNET	

Fig. 5.10 V-LAB network overview

The section titled "Operation Structure" (see Fig. 5.11) refers to the analysis viewpoint describing the operational activity of the network and then concerning not only operations and volumes of the production and distribution of goods or services, but also the logistic transport capacities over internal links and the different skills employed in the industrial network.

The section titled "Organization Structure" (as in Fig. 5.12) refers to the analysis viewpoint concerning the organization of the production and distribution activities in the network. It takes into account each enterprise in the network as an aggregated industrial body, individually managed, and so it emphasizes aspects regarding the communication of management responsibilities, the negotiation of internal control mechanism or agreements with external bodies and the effectiveness of the organization chart or coordination strategy.

The section titled "Interaction with Socio-Economic Environment" (as illustrated in Fig. 5.13) refers to the analysis viewpoint concerning the flow of products, goods demand and information with the environment in which the network lives and grows. So, in order to explain these interactions, obtaining information about negotiation or commercial agreements with clients and supplier, existing innovation programs, forecasting of dynamic evolution of the network, is mandatory. These data are collected in the CODESNET V-LAB format as shown in Fig. 5.13.

Operation Structure	
Qualitative Information	
Division of labour among firms	Indicate here the main working or servicing activities in the network. (example: in case of production, list the main activities in the production sequence, and the groups offirms which execute them; firms can be viewed as "upstream firms", "downstream firms" or "ancillary firms")
Organization of logistics and distribution	Indicate here how the logistic service is organized: if by external companies, local providers or members of the networks, etc.
Performance Indicators– Quantitative Information	
Degree of concentration of production	Indicate here the percentage of production made by the firms in the network over the whole production done in the area. In case of a very wide area, this indicator is useless.
Logistic bodies; carriers; personnel	Indicate here the logistic connections between the firms of the network, in terms of number of transportation means and storage spaces.
Applied ICT technologies	e. g. Internet connections, local networks etc.

Fig. 5.11 Operation structure (OS) data of the network presented in the V-LAB

Organization Structure	
Qualitative Information	
Agreements and control mechanisms among partners	Indicate here which types of agreements have been stipulated among the network partners to give criteria and rules for managing the partners' interactions
Responsibilities in the network	Indicate here which type of decisions for the network management a partner can take; if the case, indicate different types of partners' responsibilities.
Existence of collective agreement with external bodies	Indicate here if commercial or technical agreements exist with external institutions / companies etc., i.e. final clients and suppliers.
Existence of larger leading firms in the DESNET	
Performance Indicators–Quantitative Information	
Type of coordination body & function	Indicate here if a coordinator exists: types of coordinator body could be a committee, a managerial center, etc.
Type of organization structure	Indicate here which organization chart for the network exists, if any
Number of leading firms; average size of leading and ancillary firms	Firm size in terms of employees

Fig. 5.12 Organization arrangement (OA) data of the network presented in the V-LAB

Interaction with Socio - Economic Environment	
Qualitative Information	
Marketing strategies	Actions towards either local, or national, or European markets.
Degree of integration of the DESNET with its local labour market	Labor cost per unit time (average) and labor market constraints (for hiring/dismissal), in the local area.
Performance Indicators–Quantitative Information	
% employees in DESNET firms over the population	Percentage of employees in DESNET firms over the population (ready to work people)
Number of DESNET firms over number of firms in the area	
Note	

Fig. 5.13 Data concerning the ISEE of the V-LAB represented network

All these aspects represent the nine issues, or categories, on which the documents catalogue inside the web portal is based. For disseminating needs, all information and data required for being included in a V-LAB format have to be public, both for legal aspects and to be more sure of their reliability.

5.1.3 Virtual Library

The Library is the other main "gate" of the CODESNET website. It includes documents, called V-LIB, provided prevalently by academics, that summarize the state of the art, methods, models and case studies, resulting from researchers in the field of industrial networks. The first part of the V-LIB links both to the scientific paper summarized there and directly to the author in order to identify research groups which are studying problems related to the main analysis issues.

To allow an easy connection between a SME network described in a V-LAB document and the theoretical model/procedure contained in V-LIB papers, the V-LIB format has the same list of identification keywords as the V-LAB format.

In the following part of the V-LIB format (illustrated in Figs. 5.14 and 5.15), a concise description of the main aspects of the paper contents, of the problem approached and the results obtained is presented to provide the usefulness of the paper.

Brief description of the paper/report	
Argument	<u>Write here</u> a concise statement outlining the scope of the paper and its potential utilization (max 3 lines)
Original reference	<u>Write here:</u> Author(s) name(s) and mail address(es) Title of the scientific paper Book – journal – conference Author and Publisher mail address Year – ISBN

Fig. 5.14 Heading space of the V-LIB format

Description		
	English	*National language*
Description of the problem approached	<u>Write here</u> a description of the following items: (a) the problem approached in the paper; (b) the proposed solution and/or the approach adopted; (c) the main results; (d) the main references necessary to understand the paper content. **(max 15 lines)**	
Results	<u>Write here</u> a clear presentations of the results obtained, i.e. models, procedures, methods, and specify the potential application fields.	

Fig. 5.15 Description of the contents of the scientific paper presented in the V-LIB

After the description of the problem approached in the paper, and of the obtained results, each V-LIB includes a section of prevailing importance for the end user. As shown in Fig. 5.16, these questions are answered: (1) what's new in the paper, with respect to existing literature; (2) what's useful in terms of models to

Information for the End Users and Links	
What's new ?	Write here the novelty of the paper content and results with respect to the literature **(max 5 lines)**
What's useful ?	Write here the potential usefulness of the paper results and the industrial field of utilization **(max 5 lines)**
Tools (technologies) for the implementation of results.	Write here the methodological tools (which mathematics, languages, etc.) and the technologies (S/W and H/W) required for the implementation of the resultsobtained in the paper. **(max 5 lines)**
Links to the Main CODESNET Issues	To be compiled by the CODESNET partners Suggestions by the author(s) are welcome (please look at the Library input page for the list of 9 issue
Additional Remarks	Empty space to add your notes Max number of lines = 2

Fig. 5.16 Main information concerning the paper presented in the V-LIB format

explain, procedures to either solve or evaluate, and so on; (3) which technologies should be used to implement the paper's results.

All considerations answering the above three questions can find further explanation in the final section where a schematic summary of the paper contents can be found.

The reader can find the archive of V-LIBs, as well as the catalogue of V-LABs, on the CODESNET website. The following sections of this chapter will give further suggestions on how to use V-LIBs and V-LABs, so as to derive not only comparisons, but also significant descriptions of the weaknesses and strengths of a SME network.

5.2 CODESNET: a Bridge Between Academia and Industry

B. Caroleo and T. Taurino – Politecnico di Torino

The organization of the CODESNET website, described in Sect. 5.1 according to the paradigms of a university institute, can also have a parallel interpretation in industrial terms. Indeed, the website structure also corresponds to the typical office of an industrial analyst where, from the main entrance, the end user is immediately addressed to the list of main analysis issues, which represents the set of performance evaluation issues on which the analyst is qualified to give his/her professional support. For each main analysis issue, the end user could be addressed either towards some standard presentation of SME networks (V-LAB) with strong characteristics in terms of the issue considered, or to some presentation of scientific/technical reports (V-LIB) approaching problems concerning the same issue.

This means that the CODESNET website plays the role of a consultant who gives suggestions to an industrial end user for selecting information which can be found in some academic results. In this light the website appears to be a bridge between academia and industry, as shown in Fig. 5.17.

The nine main issues can help the analyst integrate available information about the existing industrial clusters described in the V-LAB, and models and theoretical knowledge suggested in the V-LIB.

Referring to the "bridge" between V-LIB and V-LAB illustrated in Fig. 5.17, in the first column there are the main attributes of an enterprise network that can be found in the V-LAB format, and in the third column there are the main topics explained in the collection of theoretical papers available in the V-LIB formats.

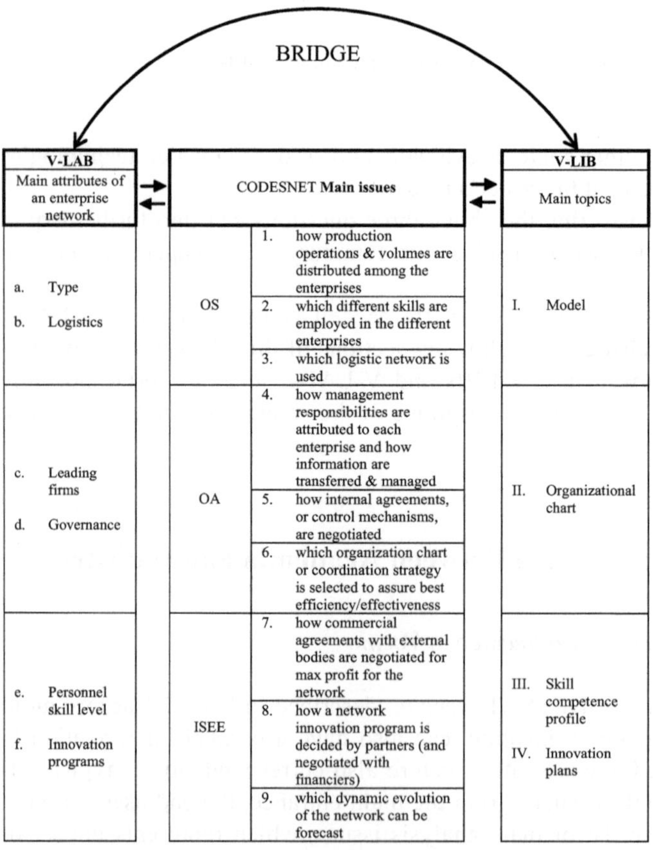

Fig. 5.17 The "bridge" between V-LIB and V-LAB formats

The central column shows the correspondences between the attributes recognized in the V-LIB and the topics of V-LAB and the CODESNET main issues are schematically presented, defined in terms of the nine main analysis issues.

Figure 5.17 gives suggestions to an end user about the utilization of the CODESNET website and its contents.

An end user, entering the website, must declare which is the main analysis issue of his/her interest. As soon as they adopt their choice, the end user can move towards the V-LAB catalogue or the V-LIB archive. In the former case, the web portal organization will present to the end user a sub-set of V-LABs, all presenting as main attributes (their strongest points) the ones directly related to the main analysis issue previously chosen, as shown in Fig. 5.17. In the latter case, the end user will receive a sub-set of V-LIBs, all concerning models and procedures to approach the selected main analysis issue. In this way, a bridge between information on interesting SME networks are connected to information on scientific/technical tools to perform the networks' analysis.

In order to explain how this bridge operates, i.e. show how the web portal can select V-LABs and V-LIBs related to a given main analysis issue, let us refer to Figs. 5.18 and 5.19.

In Fig. 5.18, for each V-LIB, the set of main analysis issues approached in the scientific paper are described. In the fourth column the type of paper content is specified; the notation introduced is (A) for algorithms and methods, (B) for case study, and (C) for survey. In the other columns, the most relevant topics approached in the paper are specified according to the list of Fig. 5.17 and the related V-LABs are listed.

From the other side, it is also important to have a reverse procedure to connect enterprises networks to one or more papers described through the V-LIB. This reverse path is shown in Fig. 5.19, where the list of the selected V-LAB formats is shown. Each V-LAB is associated to one or more CODESNET main analysis issue and is summarized by its most qualifying attributes.

These two tables describe the "bridge" between V-LAB and V-LIB lists, which are proposed to the end user when analysing elements of either the Virtual Laboratory or the Virtual Library.

Based on these tables and using the catalogue contained in the CODESNET website, some interesting data interpretation and comparison can be done, as an example of preliminary analysis which could be developed by an end user (more detailed consideration will be presented in Sect. 5.3).

First, the *numbers of enterprises* included in the different types of bodies seem to belong to specific value ranges, which characterizes each body and which could also be justified by the industrial practice. Scientific parks include more frequently from 10 to 100 enterprises, and the number is greatly dependent on the dimension of the enterprises themselves: in the case of larger enterprises, the number is lower, thus showing a specific goal or scope of the park, namely, that of promoting a balanced network of (usually) high-specialization firms. Alternatively, in the case of smaller enterprises, the role of the scientific park appears to be that of promotion and support agency. The industrial districts, network-organized, range from ten to

#	selected V-LIB list	Authors	Main Issue	Type of paper	Most relevant TOPICS: I. model	II. Organizational chart	III. Skill competence profile	IV. Innovation plans	Related V-LAB
4	The AMBIT model for strategic decision making and performance	Jagdev H., Brennan A., Browne J.	8	C				X	15 , 26 , 35 , 53 , 54 , 58 , 59 , 67
5	Analysing business information systems using integrated business models and user requirements	Shen H., Wall B., Zaremba M., Chen Y., Browne J.	4	A					8 , 65 , 69
8	Capacity allocation in distributed enterprises	Bruccoleri M., Lo Nigro G., Federico F., Noto La Diega S., Perrone G.	5 , 9	A	X	X			13 , 27 , 64
9	Collaborative networks: a new scientific discipline	Camarinha-Matos L. M., Afsarmanesh H.	4 , 6	C		X			7 , 8 , 29 , 35 , 64 , 65 , 69
12	Competence Profiling and Problem Solving in Virtual Networks	Edelmann C., Wagner K.	2	C			X		7 , 9 , 13 , 29 , 38 , 53 , 54 , 66
13	Constructing a typology for networks of firms based on activities complementarity and competences similarity	Burlat P., Besombes B., Deslandres V.	1	B			X		29 , 37 , 38 , 50 , 53 , 54 , 62 , 65 , 66 , 68
16	Contract Costing in Outsourcing Enterprises: The Role of Simulation	Liston P., Byrne P. J., Heavey C., Brown L.	7	B	X				
19	Decentralised Planning tools for Reconfigurable Enterprises	Argoneto P., Bruccoleri M., Lo Nigro G., Perrone G., Noto La Diega S., Renna P., Sudhoff W.	1 , 5 , 6	A	X	X			13 , 27 , 28 , 50
20	A decision support approach to select pertinent scenarios for collaborative competence development within a network of firms	Boucher X., Lebureau E.	2	A		X	X		13 , 29 , 37 , 38 , 53 , 54 , 62 , 64 , 66 , 67 , 68
22	Delivery of raw material to production firms with deterministic production times	Bubnicki Z., Wolkowinski K.	3	A	X				5 , 27
26	Electronic Product recovery -understanding resource recovery as a business objective	Goggin K., Browne J.	4 , 5	B					5 , 13
27	Empirical research opportunities in reverse supply chains	Prahinski C., Kocabasoglu C.	1	C					27 , 37 , 50 , 58 , 59
32	Framework for outsourcing manufacturing: strategic and operational implications	Momme J.	5 , 7	A	X				13 , 34 , 63 , 66
33	A framework for comparing outsourcing strategies in multi-layered supply chains	Abdel-Malek L., Kullpattaranitun T.	5 , 7	A	X				13 , 34 , 66 , 68
34	Generalized newsboy model to compute the optimal planned lead times in assembly systems	Ould Louly M. A., Dolgui A.	1	A					5 , 7 , 37
35	How to achieve supply chain coordination and increase profits with the use of quantity discount policy	Li J., Liu L.	6	A		X			7 , 8 , 27 , 64
36	Innovation paths in enterprises clusters	Bielli M.	8	B				X	7 , 9 , 15 , 26 , 34 , 35 , 40 , 41 , 42 , 58 , 59 , 63, 67

Fig. 5.18 Selected list of V-LIB formats, with the most relevant topics

#	selected V-LAB list	Source	Main Issue	Attributes (a) Type	(b) Logistics	(c) Leading firms	(d) Governance	(e) Personnel skill level	(f) Innovation programs	Related V-LIB
50	The Industrial Symbiosis	UNOTT	1 , 4 , 9	network 5 enterprises	material and energy exchanges	5 large enterprises				13 , 19 , 27 , 49
53	Walbrzych Special Economic Zone "Invest Park" Ltd	ILW Poznan	2 , 4 , 8 , 9	Scientific park 65 firms (all sizes)	external logistics		Support agency	importance of high competence, skills		4 , 12 , 13 , 20 , 52 , 76 , 86
54	The Suwalki Special Economic Zone J.S.C.	ILW Poznan	2 , 4 , 8 , 9	Scientific park 55 firms	external logistics	5 large firms	cooperation with banks and joint-stock company	importance of high competence, skills		4 , 12 , 13 , 20 , 52 , 76 , 86
58	ENVIRONMENT PARK SPA	TECNORETE	8 , 9	Scientific park				importance of high competence, skills	innovation; R&D	4 , 27 , 36 , 65 , 81 , 103 , 107
59	PST S.p.A.	TECNORETE	8 , 9	Scientific park 25 companies				importance of high competence, skills	innovation; R&D; collaboration with universities	4 , 27 , 36 , 65 , 81 , 103 , 107
62	COMO TEXTILE DISTRICT	POLITO	1 , 2 , 5 , 8	SC 10 SMEs		2 large firms		importance of high competence, skills	innovation; R&D; collaboration with universities	13 , 20 , 72 , 77
63	INDUSTRIAL TANNING DISTRICT	POLITO	5 , 8	Multi-stage SC 900 SMEs						32 , 36 , 72
64	WEAVING DISTRICT OF PRATO	POLITO	2 , 5 , 6	Multi-stage SC 7,273 SMEs	semi-processed and finished products exchanges		District Committee	several different skills		8 , 9 , 20 , 35 , 56 , 101
65	LECCO TEXTILE DISTRICT	POLITO	1 , 4	network 116 enterprises		30 large firms	District Committee	importance of high competence, skills		9 , 13 , 5 , 69 , 84
66	INOX VALLEY	POLITO	1 , 2 , 5 , 8	network 1000 enterprises	external logistics			several different skills	R&D	12 , 13 , 20 , 32 , 33 , 89
67	SASSUOLO TILES CERAMIC DISTRICT	POLITO	2 , 7 , 8	network 199 enterprises	external logistics	5 large firms			innovation	4 , 20 , 36 , 72
68	AGRICULTURAL DISTRICT: NOCERA - GRAGNANO	POLITO	1 , 2, 5 , 7	SC 200 SMEs	external logistics		District Committee	several different skills		13 , 20 , 33 , 69
69	NAUTICAL AND MECHANICAL SHIPBUILDING	POLITO	1 , 4 , 6				District Committee			5 , 9 , 84

Fig. 5.19 Selected list of V-LAB formats, with the most qualifying attributes

one or two hundred component firms. This pattern of firm groupings, however, does not seem to be sufficiently motivated: it could be a characteristic of the set of networks collected by the CODESNET project. A clear pattern occurs for supply chains, which ranges from hundreds to thousands of components. This is a typical organization in several supply chains operating in a specific industrial sector producing mass customized products (as for automotive or electronic goods). Regarding development agencies, these last ones are a specific French organization, and only few very similar data have been available to the CODESNET project. However, their main features are really interesting for a comparison with the scientific park, with which they share a similar mission.

According to another viewpoint, the types of governance have been surveyed: it has been classified either as cluster/network committees, or with the presence of a leading firm, or of a support/coordination agency, or through individual agreements. Considering this second characteristic, the differences among types of analysed industrial bodies appear to be more recognisable.

Among the considered industrial districts/networks, the presence of a committee (sometimes only political, but also managerial) is frequent: this feature has also been detected in a number of other Italian districts of similar type (independently analysed for a parallel scientific research project developed at Politecnico di Torino in cooperation with the regional government). A smaller number of districts presents a small group of leading firms: usually this pattern can be found in districts producing customized goods (e.g. jewellery), where the leaders are firms which also characterize the district trademark. Presence of individual agreements linking the component firms rarely occurs in the set of analysed districts. Considering the supply chains, both the presence of coordination committees, and of support agencies and leading firms can be found: no special type of governance can be recognized. However, a different situation between a supply chain managed by either a committee or a support agency, on one side, and that controlled by a leading firm, is related to the type of production and the dimensions of enterprises included: in the former case, firms of similar dimensions agree in having a coordination structure for managing the production flows as well as the interactions with markets; in the latter, the leader drives all parts of the chain, being usually the enterprise at the end of the chain and then supplying final products.

In the case of scientific parks, a prevalence of support agencies can be seen, with respect to presence of leaders. The first situation is the usual one both in Italian parks, and in the French "poles of competitiveness". Finally, in the few development agencies detected here, only coordination committees are used.

Good suggestions can be obtained also by considering the *requested skills of personnel employed* in the analysed industrial bodies. A variety of skill types, i.e. high-level skills, or specialized skills, or many skills without any special request, can be found both in the supply chains and in the industrial districts: this sparse pattern depends on the usual operation structures of these two organizations, that means lines or networks for mass production of consumer goods, calling for any type of skill, from workers to designers (a few) and to managers. On the contrary, scientific parks only call for high-level skills, thus justifying their main function of

Table 5.1 Innovation of industrial bodies

Innovation	Supply chains	Districts/networks	Scientific parks	Support agencies
Yes	27%	33%	70%	N.A.
No	73%	66%	30%	N.A.

promotion and support agency already recognized by considering both the number of enclosed firms and the usual types of governance.

Finally, other complementary good considerations can be done by summarizing the numbers of different industrial bodies, which started or did not start *innovation programs*. The percentages can be seen in the small (but interesting) Table 5.1.

The last data, referred to as support agencies collected by the CODESNET project unfortunately, is too small to be significant. For the other types of networks, the estimated percentage of industrial bodies launching innovation programs or having innovation actions under development give a clear answer to the initial question: is the action of supporting chains and networks a good social-political strategy for Europe? The answer now is evident: these industrial bodies, not yet equipped with a robust organization for an effective governance, but so widely diffused in the European industrial world, need to be supported if their survival in the worldwide markets is desired, as it must be.

5.3 Analysing and Comparing SME Networks Through the CODESNET Website

D. Antonelli and T. Taurino – Politecnico di Torino

In this section two examples are given of the usability of the CODESNET website for comparing and benchmarking of networks.

The main point that needs to be stressed is that it is of no use to score different networks all over the Europe, using arbitrary indexes, and forming an ordered list with winners and losers. It is much more proficient to state a point of view from which to observe the networks and to try to extract the key factors that influence the network behaviour from that point of view.

Furthermore it must be aware of the fact that data are not disposable in a quantitative numerical format and therefore it is hazardous to use them for scoring.

The information and the data for each cluster are contained in the standardized form called V-LAB, as described in Sect. 5.1. Each V-LAB gives a detailed description of the main qualitative attributes that best illustrate the status of the ID. It also contains an analysis of quantitative key performance indicators. This information gives a complete report on the operation structure, the organizational arrangement and the relation with the socio-economic environment.

In this section, the first example is a SWOT analysis applied to a selected number of networks in order to evaluate the potential of collaboration; the second example is a statistical analysis of the factors that influence innovation inside the networks and is applied to a larger number of homogeneous networks.

5.3.1 SWOT Analysis of the Networks[1]

SWOT analysis (Fig. 5.20), is a strategic planning tool used to evaluate the *strengths, weaknesses, opportunities,* and *threats* involved in a project or in a business venture. It involves specifying the objective of the business venture or project and identifying the internal and external factors that are favourable and unfavourable to achieving that objective. The technique is credited to Albert Humphrey, who led a research project at Stanford University in the 1960s and 1970s using data from Fortune 500 companies (Wikipedia 2008).

The comparison hinges on the three criteria used to organize the data of the districts examined. The ensuing quality assessment of the benchmark district is a key tool to: point out the strength and weakness of the CODESNET website compared to that of other districts, and suggest the improvements required to ensure a better usability by all partners involved, whether the districts or a third party intends to do research in industrial districts.

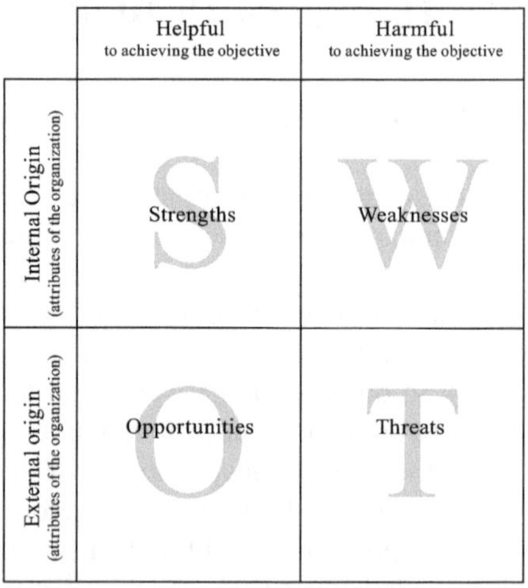

Fig. 5.20 SWOT analysis scheme

[1] Part of this section is the result of a test utilization of the CODESNET website performed during the ALFORM 2007 Master Course at the University of Trieste. The authors wish to thank the Master Course students: L. Coren, P. Mongini, E. Muriglia, D. Naresi, S. Simioni, M. Tomas, E. Toppano, and Prof. W. Ukovich, Master Course coordinator.

The comparison concerns the following districts:

1. The Cork district, located in the province of Sassari, processing cork for semi-manufactured and final products
2. The GIER district, in the French region of the Loire, providing services related to: production, manufacture, engineering, planning, financing, administration, cooperation, communication, coordination and logistics
3. ICSE – ZNTK, a Polish district where locomotives and rolling stock are repaired and maintained
4. The Spectacles district in the province of Belluno (benchmark district)

The data essential to the comparison are homogeneously listed in Table 5.2.

Unlike three districts with an exclusively local field of action, the ZNTK district, which cooperates with Polish Railways (PKP), is active nationwide.

Table 5.2 Comparison of four industrial districts

	Spectacles District	Cork District	GIER	ZNTK
		OPERATIONAL STRUCTURE		
Number of Firms	560	130	120	2(+172)
Division of Labour among firms	Sub-supplying relationship	Sub-supplying relationship and product differentiation	Different expertise and experience; encourage inter-enterprise cooperation	ZNTK Olenisca: locomotives ZNTK Olawa: cars, tank cars
Organization of logistics and distribution	Common distribution platform: Out-and-Out Supply Chain Management	N/A	External bodies and club memberships	Internal and external Transport
Logistic bodies carriers: personnel	Logistic network is that related to the five leaders	N/A	N/A	Own and external logistic bodies are used
		GOVERNANCE		
Agreements and control mechanisms among partners	"Sipao" main point of reference for the administrations	No definite agreements among enterprises	No general agreement	Cooperation agreements
Responsibilities in the network	Leader firm affect organization and production of the whole district	No precise responsibility	A team of three members animates the club	Responsibilities follow law regulations and rules of cooperation in agreements
Existence of collective agreement with external bodies	Padova University, designers, industrial association, chamber of commerce	No collective agreements	Stimulating the local economy actors through financial and consulting partners, institutions	Only individual agreements
Type of coordination body and function	A committee to coordinate activities of institution and category association	There is not a district committee	The club is managed by: • 1 chairman • 5 committees	No coordination body
		INTERACTION WITH SOCIO-ECONOMIC SECTOR		
Degree of interaction of the network with its local labour market	Integration degree with local labour market is very high	High integration degree	Closely integrated	Permanent integration with local labour market
% employees in network firms over the population	About 13% of employees of the Province of Belluno work in the District	8,6%	About 10% of local ready to work people	733 persons

The Cork, Spectacles and GIER districts are deeply rooted in their regions, employing about 10% of the local population, and are composed of small and medium concerns. However, the ZNTK, despite their 733 employees, has no significant advantage at a national level.

The difference between the afore-mentioned districts lies in their organization. With the exception of the Spectacles district, the others boast no hierarchy, but rather a number of cooperation agreements between their companies: cooperation without formal agreements for companies of the Cork district; cooperation agreements based on international technical standards (ISO) for the ZNTK; in the Loire district (GIER) the involved partners aim to use a regional trademark for their products in international fairs and exhibitions. As to the Spectacles district, five leading companies define the guidelines related to style and quality and have a shared supply chain. Subordinate companies use in turn high-quality standards in order to withstand global competition and solely interact with the leading concerns. The hallmark of this district is a range of agreements with regional research and training centres in the field (e.g. University of Padua). Moreover, their trademark and fame as well as their agreements with great designers show how considerable their market share is.

The Cork district meets most domestic and, partially, foreign demand. Over the last few years, the export decrease has been indicative of difficulties due both to insufficient cooperation inside the district and to the small size of the companies, which are unable to withstand international competition.

ZNTK accounts for 10% of the market and beat the competition of 11 companies in the field. No data are available on GIER's market share.

The assessment of the benchmark district focuses on the analysis of the information related to organizational arrangement (governance), operational structure, and the interaction at a socio-economic level.

The data heterogeneity hinders the analysis at a quantitative level. Therefore quality is the sole yardstick used to assess districts according to the aforementioned information categories in compliance with the following test. A graph with the judgment values for each category will help carry out a SWOT analysis of the districts. There are four levels for each information category (Table 5.3).

As a result of the analysis, levels ("votes") of each district for each category are summarized in Table 5.4.

Figure 5.21 shows that the Spectacles district, with a medium level for each category, is the most balanced. However, the district lacks in governance and operational structure compared to the ZNTK, gives less advantage to the regional socio-economic sector in comparison with the Cork district and beats the competition of the GIER district.

Table 5.3 Values measuring the levels of organization, and interactions with the outside

	Level of organization in the operational structure
0	No organization or communication between companies
1	Low level No centralized logistic network informal agreements
2	Medium level Incomplete logistic coordination only at certain production stages
3	High level Logistic management in and outside the district, shared staff management
	Level of governance
0	No district body
1	Low level Representative bodies
2	Medium level Advisory structures aimed at the information diffusion and at the development of the district
3	High level Complex structures for the action planning according to specific provisions and/or constraints
	Level of interaction with the socio-economic sector
0	No interaction Local people employed less than 2% No district Incompatible with the regional socio-economic situation
1	Low interaction Local employment rate 2 to 5% No district
2	Medium interaction Employment 5 to 10% Second generation district
3	High interaction Local people employed >10% Old district Training centres linked to the district

Table 5.4 Votes concerning governance, structure and interactions with the outside for the evaluated four districts

	Gov	OS	ISEE
GIER	2	1	2
ZNTK	3	3	2
Cork	0	1	3
Spectacles	2	2	2

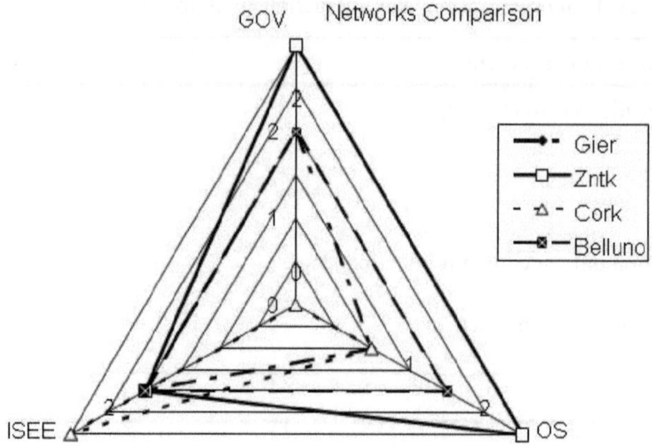

Fig. 5.21 Spectacles district balance of network components

5.3.2 Principal Component Analysis of the Innovation Level Inside SME Networks

The central idea of the principal component analysis (PCA) is to reduce dimensionality of a set of data with a large number of interrelated variables, while retaining as much as possible the variations. To achieve this objective, PCA transforms the data to a new set of variables, the principal components (PCs) (Jollife 2004). PCs are uncorrelated, and are ordered so that the first few components retain most of the variation present in all of the original variables (Gifi 1981). Data consists of qualitative or categorical variables as well as numerical variables that describe the indicators for a limited number of categories. The zero in these scales is uncertain, the relationship among the different categories is unknown, and although some of the variables are composed of categories that are ordered, their mutual distances are still unknown. The uncertainty in the unit of measurement is not just a matter of measurement error because its variability may have a systematic component (Gifi 1985).

An important development in multi-dimensional data analysis has been the optimal assignment of quantitative values to qualitative scales (Meulman et al. 2004). This kind of optimal scaling is a very general approach to deal with multivariate categorical data. The optimal scaling process turns qualitative variables into quantitative ones. Optimality is a relative notion because it depends on the particular data set that is analysed. The non-linear optimal scaling transformations of ordered categorical or continuous ordinal data are handled by means of monotonic transformations, maintaining the order in the original data. Categorical or otherwise nominal data in which the categories are not ordered will be given an optimal quantification, called scoring. Non-monotonic functions can also be used for continuous numeric and ordinal variables when non-linear relationships among the variables are assumed. In these cases, it is often useful to collapse the data in a limited number of categories (sometimes called binning) and find an optimal quantification for the categories.

The analysed data come from 20 industrial districts (IDs) distributed on different European countries, namely Italy, France, Germany, Great Britain, Greece and Poland, catalogued in the CODESNET Web Portal. At the current level of the database, only a sub-set of the existing IDs is described as a V-LAB.

The advancement process in innovation can be described by the three categories of indicators proposed by Lipparini and Lorenzoni (1996):

- Learning by specializing
- Learning by localizing
- Collective learning

Learning by specializing is derived from the flexible specialization approach (Amin 1991) and from the industrial divide theory. The main characteristics of this kind of innovation are found in the focus towards the resources and skills involved in the ID, as they are the central core of any enterprise and indeed of the ID seen as an extended enterprise. It also concentrates on the capacity to rapidly adjust to demand changes, thanks to the flexibility and specialization of the district. The latter being solved thanks to the skills enclosed in the district and the former being explained by the ICT used.

Learning by localizing comes from the classic district approach that follows Marshall's economic theories (Amin 1991; Maillat et al. 1993). The key term in the dynamics of learning by localizing is Marshall's industrial atmosphere theory. It highlights the unintentional characteristic of the relations between companies and also the importance of the local environment and atmosphere which can exclude any company that does not take part in it. This is how the learning process can become a competitive advantage with respect to companies not located in the ID.

Collective learning derives from the studies of the GREMI (Groupe de recherche européen sur les milieux innovateurs) (Ratti et al. 1997) in which the local system organically reduces uncertainty in the innovation processes, it creates a web of interaction, it generates conventions and behaviour precepts and shared codes of inclusion and exclusion from the ID.

5.3.2.1 Learning by Specializing

The process is described by variables (Table 5.5) that explain what kind of resources or industrial functions are included in the district, what level of ICT has been implemented and variables that describe the dimension of the cluster in terms of how large the production turnover is on average, how many workers are involved in each districts and also a value of the average size of firms.

Once the variables are defined, the analysis can be executed just as in linear principal component. The first step is to verify that each variable is at least mildly correlated with the other variables. If there are any variables that are not correlated

Table 5.5 Learning by specializing: variables

Learning by Specializing		
Variable	**Description**	**Metric**
Division of labour	Labour distributed among the network	Number
Total Labour	Dimension of the network in terms of Labour	Number
Skill	Skill involved in the network	Vector
Level of production	Dimension of the network in terms of production or the amount of investment	Number
Logistics' quality	Type of logistics involved	List
ICT	ICT distributed in the network	Vector

with others, they should be left outside the PCA. The correlation check is made through the analysis of the correlation matrix.

The next step is deciding how many components should be used to best describe the variables. The problem is solved by initially analysing the data with a number of dimensions equal to the number of variables. Using the equivalence results in having a plot in which each dimension is described by a variable. As consecutive dimensions are extracted each one accounts for less and less variability. The decision of when to stop extracting factors basically depends on when there is a small variability left. The nature of this decision is arbitrary; however, various guidelines have been developed. The ones used in this research are called the *Kaiser criterion* and the *Scree test*. They are applied together to force a more reliable decision. The Kaiser criterion says that only dimensions with eigenvalues greater than or equal to 1 must be retained. In essence this is like saying that, unless a factor explains at least as much variance as the equivalent of one variable, it should be left out. This criterion was proposed by Kaiser in 1960 but it is still one of the most used methods. The second method is the Scree test, and because it is a graphical method, allows to generate an estimate of the trend with which the eigenvalues change with increasing number of dimensions. In the present case the best number of dimensions is two as summarized in Table 5.6.

Table 5.6 Learning by specializing: correlation of transformed variables

	Division Labour	Tot Labour	Skills	Production	Logistics	ICT
Division Labour	1.000	0.635	0.012	0.235	–0.059	–0.245
Tot_Labour	0.635	1.000	–0.019	0.628	–0.290	–0.185
Skills	0.012	–0.019	1.000	–0.059	0.292	–0.580
Production	0.235	0.628	–0.059	1.000	–0.261	–0.137
Logistics	–0.059	–0.290	0.292	–0.261	1.000	–0.235
ICT	–0.245	–0.185	–0.580	–0.137	–0.235	1.000
Dimension	1	2	3	4	5	6
Eigenvalue	2.185	1.783	0.811	0.635	0.382	0.204

Table 5.7 Learning by specializing: component loadings

	Dimension	
	1	2
Division_Labour	0.777	0.114
Tot_Labour	0.982	–0.074
Skills	0.012	0.888
Production	0.792	–0.124
Logistics	–0.367	0.584
ICT	–0.333	–0.855

After having verified how many dimensions are necessary in the PCA, the first information obtained from the output of the analysis is the component loading table (Table 5.7), which can be interpreted by comparing the values for each variable on every dimension.

The result of this simple observation is that dimension 1 mainly explains *Division_Labour, Tot_Labour* and *Production*, whilst for dimension 2 the variables that can be projected are *Skills*, *ICT* and *Logistics*. The variables are well-grouped on each dimension and are almost orthogonal to the same directions as the dimensions.

From Fig. 5.22 it is apparent that skills employed in the district are positioned almost on the vertical zero, which undoubtedly results in assigning it to dimension 2, on the positive side. Conversely, ICT is explained mainly by dimension 2, on the negative side. The larger component stands on dimension 2. The same can be said about logistics, which is found in the same direction as skills. The difference here is that logistics is not as strong in explaining dimension 2 as skills is, but still gives a better understanding of the result. On dimension 1 other three variables are present, but all of them are co-linear and also close to the horizontal zero. In this second case, if the variables are analysed carefully a possible definition for dimension 1 can be the term *district dimension*, as the variables projected on it give a description, in different terms, of the size of the industrial district. On the other

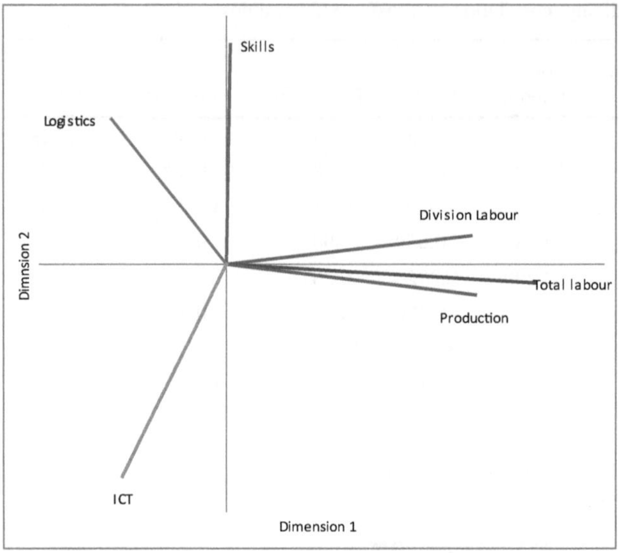

Fig. 5.22 Learning by specializing: component loading plot

Table 5.8 Learning by specializing: some ID scores

ID	Dim	Skills	ID	Dim	Skills
1	–0.178	–0.046	8	–0.447	–0.497
2	0.573	–1.599	9	–0.126	0.730
3	4.199	0.051	10	0.376	0.867
4	–0.445	–1.473	11	–0.289	–0.013
5	–0.146	–1.516	12	0.272	0.162
6	–0.603	0.962	13	0.018	–0.920
7	0.276	0.536	14	–0.109	1.097

axis the main focus is on the skills involved in the district, that are characterized by the industrial functions involved and the ICT tools used. The proposed name for dimension 2 is then *district skills*. Scores for IDs are shown in Table 5.8.

Each ID is characterized by two values which identify it relatively to the two dimensions found in the previous step. A more functional way to see this result is by using the biplot on which both the variables and the object are shown.

The biplot shows each district according to its most significant characteristics (Fig. 5.23). The position of the IDs illustrates how the more important discriminating factor lies on the vertical axis, as on the horizontal one there is a very low distribution.

What actually differentiates the clusters is the difference in skills employed in the SMEs. Exceptions are a few IDs that are positioned on the far right of the plot. These outliers have such an important explanation on the dimension axis because they are large supply chains, for which the describing parameters of the dimensions

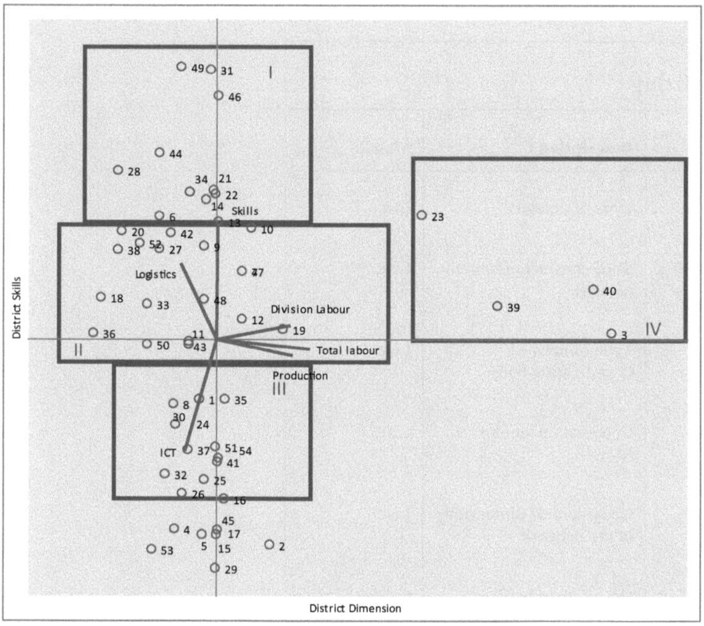

Fig. **5.23** Learning by specializing: biplot

are a few orders of magnitude larger than any other district. The final step of the analysis of the results is to create groups among the clusters. This is done by observing the biplot and by creating clusters of IDs that have similar characteristics. For instance, the outliers on the far right have similar dimensional qualities. In this case the districts have chosen to innovate by using a magnitude approach, so the larger they are, the better chance of innovating. This result gives a good idea of how a large number of districts can be subdivided into a small number of types.

5.3.2.2 Learning by Localizing

This system looks at how IDs localized in the territory as districts that are distributed on a very large scale occur more frequently.

There are two types of possible occurrences: one being the presence of SMEs of similar size and the other being the exact opposite, where the presence of a predominant enterprise pushes the district towards its own goal.

Another attribute used to describe this learning system is the relation the ID has with external entities, mainly due to the economic relations linking the enterprises.

To find the number of dimensions necessary to obtain a good use of the variables a full analysis, adopting the same number of dimensions as of variables, is done (Table 5.9). After having chosen the number of dimensions, the correlation analysis is repeated with two dimensions.

Table 5.9 Learning by localizing: variables

Learning by Localizing		
Variable	**Description**	**Metric**
Structure model	Type of network	List
Relative importance of SMEs	SME consideration in the network	List
Coordination	Type of network Coordinating body	List
Cooperation	Cooperation among SMEs	List
Distribution of the territory	Geographical dimension of the network	List
External Relations	Openness to bodies external to the network	Grade
Export	Network international relationships	Number
Size of district	Number of SMEs in the network	Number

Dimension 1 is defined by the governance type so it is labelled *Governance*, remembering that a positive value means that the main concern is cooperation, whilst a negative value indicates coordination. Dimension 2 is described mainly either by the market radius or by the existence of leaders in the SME so it can be labelled *Leading to external*, where a positive value indicates a leader and large global market and vice versa on the negative side of the axis.

The biplot (Fig. 5.24) shows a very dispersed graph due to the choice of two dimensions. The obvious suggestion is that three could be a better number for the space dimensions, but for sake of simplicity we have chosen to project the space on a plane. It is interesting to observe that the IDs have very different approaches to learning by localizing. It is actually difficult do define similar groups by observing the graphs as they are so dispersed but nonetheless a few clusters are definable. Two major groups (I and IV) are the ones defined along the governance dimension, as shown in Fig. 5.25; on the negative side IDs with a main coordination objective can be found and conversely on the positive side IDs interested in cooperating with each other. Other groups that have been found among the data are II and VI which respectively indicate IDs with equally important SMEs and IDs in which at least a leading enterprise can be found. Group III is instead de-

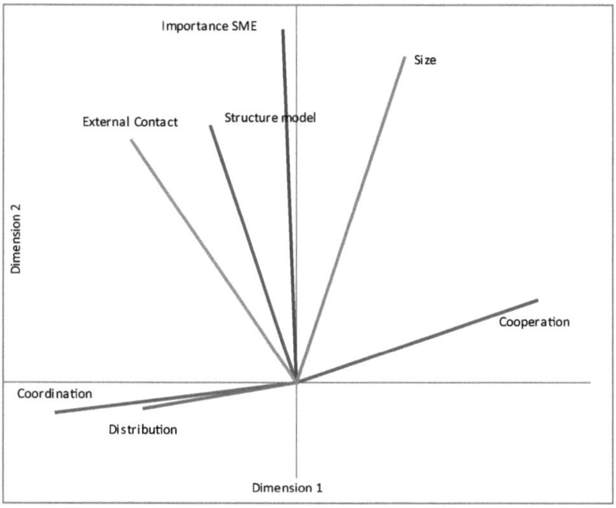

Fig. 5.24 Learning by localizing: component plot

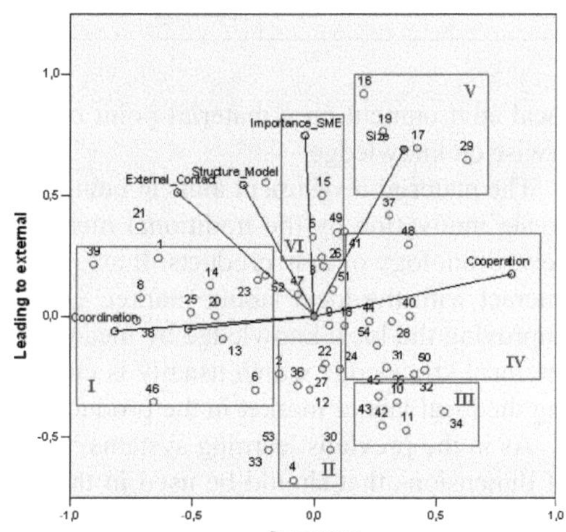

Fig. 5.25 Learning by localizing: biplot

fined by the structure model, or in other words the type of ID, which, as it is shown in the biplot, is primarily a group of virtual networks. Finally group V contains a few outliers as they all are particularly large industrial districts in terms of number of SMEs present.

5.3.2.3 Collective Learning

Collective learning is the last learning system considered (Table 5.10); it is probably most connected to the culture of the district and how the district invests in the

Table 5.10 Collective learning: variables

Collective Learning		
Variable	**Description**	**Metric**
Investments in R&D	Interest in innovating using R&D programs	Grade
Labour Market	Network connection to the local population	Grade
Degree of communication	Grade of communication inside the network and towards external bodies	Grade
Vision capability	Network preparation for future challenges	Grade
Level of competitiveness	Readiness of the network	Grade
Capability to transfer knowledge	Grade of partnership through transfer of knowledge	Grade

local environment on a material point of view or also on an immaterial one, otherwise on knowledge.

The material investment aims at being the one where research and development create innovation by the traditional means of investing money in the research of new technology or new products. It also includes how the enterprises of the cluster interact with the local labour market. Often mission statements of IDs consist in improving the local knowledge by means of innovating by R&D but also by using the local knowledge which usually is tacit, and it can only be absorbed by including the local labour market in the production process.

As in the previous learning systems, the first step is to find the optimal number of dimensions that should be used in the analysis. The correct number of dimensions that should be used in the analysis is again 2, as only the first and second eigenvalues are larger than one, which is the theory behind the Kaiser criterion.

Observing the component-loading plot of Fig. 5.26 it is possible to see that two groups of variables seem to be somewhat orthogonal to each other, but not along the dimensions of the pure principal component analysis. By a simple rotation the variables could be projected onto two dimensions in order to find a clear projection on the dimensions. The rotation of the axis is performed by an orthogonal matrix that maximizes the components of the variables on the dimensions. This does not change the result of the analysis as in the process also the objects are rotated using the same transformation matrix.

The result of the rotation of the variables is shown in Fig. 5.27. The variables *Investments* and *Labour_Market* can be projected onto dimension 2, whilst the remaining variables, all the variables that discuss the knowledge factor of the ID,

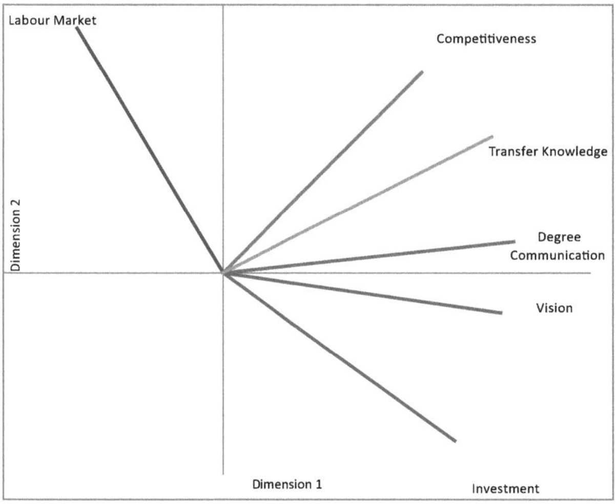

Fig. 5.26 Collective learning: component loadings plot

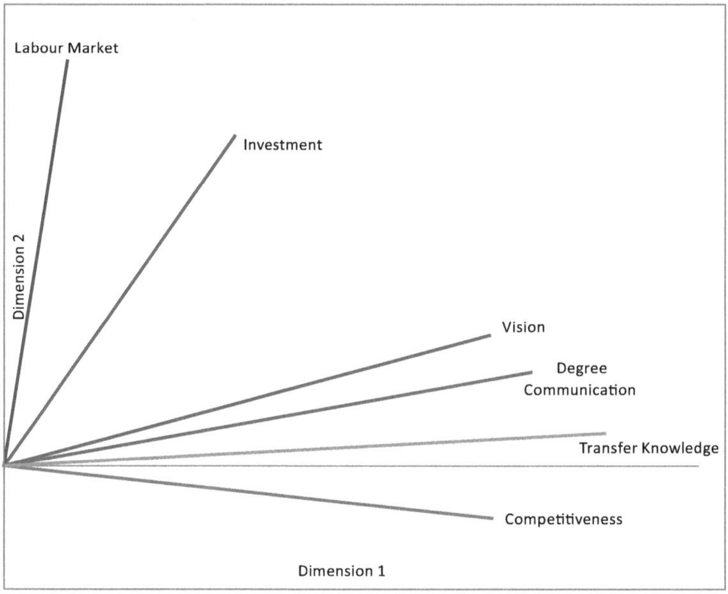

Fig. 5.27 Collective learning: named dimensions plot

are positioned very close to dimension 1 on the positive side. It is true that investment is not exactly vertical, which means that a small part of dimension 1 can also be explained by using this variable, but the predominant projection of it is on the vertical axis. Just like in the previous learning systems the interest at this point is to be able to name the dimensions according to the variables that lie on it. The key word that describes the variables on the vertical axis is investment as there is an *investment in R&D* variable and an *investment in Labour_Market* variable. Dimension 2 then can be called *territorial investments*. On the horizontal axis the main issues that are described are the knowledge factor, as all the variables indicate how much knowledge is being gained or transferred or, in any case, produced and used. So the horizontal axis can be called *knowledge factor*.

The biplot in Fig. 5.28 shows each district according to its most significant characteristic and is presented with the groups outlined and numbered, according to the different features that can be found.

The position of the ID are very scattered to show that there are major differences between the IDs on the collective learning scale. The larger amount of objects are positioned either close to the knowledge factor or around the investment in the labour market. This is understandable as in the short term, the production coefficients have to be considered fixed and so the large monetary investments in research and development are not as common as investing in knowledge factors and in the local labour market.

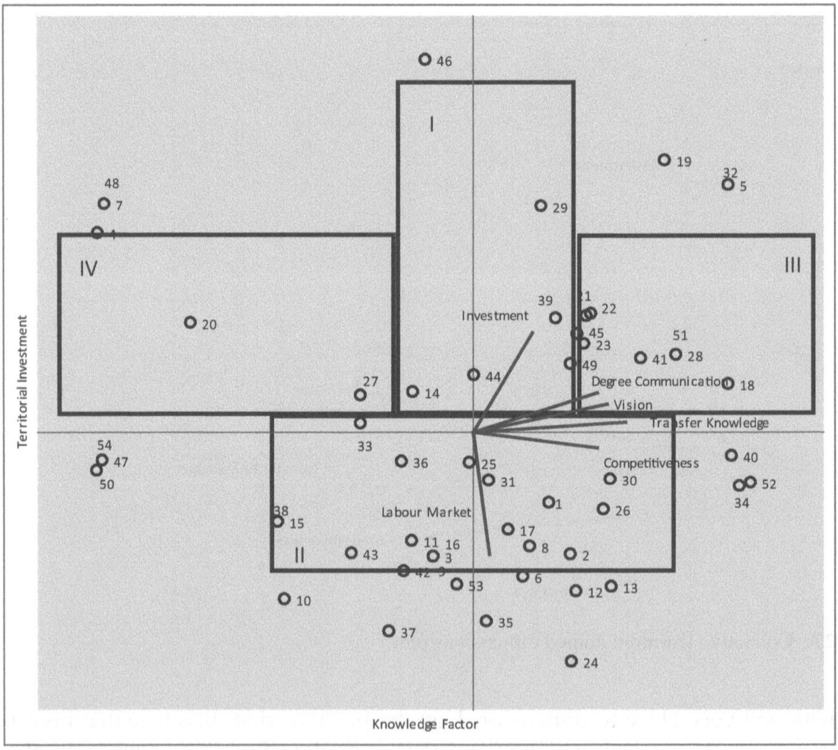

Fig. 5.28 Collective learning: biplot

References

Amin A (1991) These are not marshallian times. In: Innovation networks and spatial perspectives. Bellhaven, London, pp. 105–117

Gifi A (1981) Nonlinear multivariate analysis. Leiden University, Department of Data Theory, Leiden

Gifi A (1985) PRINCALS. Internal Report UG-85–02, Leiden University, Department of Data Theory, Leiden

Jolliffe I (2004) Principal Component Analysis. Springer, Berlin Heidelberg New York

Lipparini A, Lorenzoni G (1996) Le organizzazioni ad alta intensità relazionale. Riflessioni sui processi di learning by interacting nelle aree ad alta concentrazione di imprese. "L'Industria", a. XVII(4):817–839

Maillat D, Quevit M, Senn L (1993) Reseaux d'innovation et mileux innovateurs: un pari pour le dévelopment régional, 1, Neuchatel

Meulman J, Van der Kooij A, Heiser W (2004) Principal Component analysis with nonlinear optimal scaling transformation for ordinal and nominal data. In: Kaplan D (ed) Handbook for quantitative methods in the social sciences. SAGE, Newbury Park, CA

Ratti R, Bramanti A, Gordon R (1997) The dynamic of innovative regions: the Gremi approach. Avebury, Aldershot

Wikipedia: The Free Encyclopedia (2008). http://en.wikipedia.org/wiki/Graph_theory. Accessed 2008

Authors

Stamatiki AGOTI received her MSc in Economics with the title "Applied Microeconomics and Marketing" in 2006 from the University of Patras. Currently, she is working in Patras Science Park, department of Project Planning and Implementation. Among her responsibilities is also the management of GSRT projects.

Dario ANTONELLI Italy 1966, holds a MSc degree in Mechanical Engineering from the Politecnico di Torino (1990). He worked at Fiat Research Center until 1992. He is currently Associate Professor at the Department of Production Systems and Economics of Politecnico di Torino, and he is enrolled in the Faculty of Management and Industrial Engineering. His scientific activity is mainly related to numeric finite element simulation of metal working processes, to experimental identification of process parameters and to supply chain management. He is presently responsible for the research project on micro-assembly under the Italian Ministry grant.

Miryam BARAD was Head of the Industrial Engineering Department at Tel Aviv University and held academic visiting positions at universities around the world.

Her research and professional interests focus on various aspects of quality and flexibility. Prof. Barad is on the Editorial Board of the *International Journal of Production Research*, a Council member of the International Foundation of Production Research and a member of the New York Academy of Sciences.

Xavier BOUCHER is currently Associate Professor in Industrial Management at the École Nationale Supérieure des Mines de Saint Étienne (France). His current research focuses on virtual organizations, agile logistic systems and decision models to manage the supply chain agility. His previous research works on integration of socio-technical systems led him to publish in several journals like: *International Journal of Computer Integrated Systems*, *Computers in Industry*, *Production Planning and Control*, *Robotics and Computer Integrated Manufacturing*, and *Concurrent Engineering*.

PJ BYRNE is a Lecturer of Operations/Supply Chain Management in Dublin City University Business School, having formerly worked as a Senior Research Fellow at the University of Limerick. He received a PhD from the Manufacturing and Operations Engineering Department at the University of Limerick, in the area of supply chain simulation. His research interests are supply chain design, analysis and optimization, environmental impacts of supply chain construction, company outsourcing, decision-making and costing, and industrial simulation applications.

Brunella CAROLEO was born April 14, 1981. She graduated in Mathematical Engineering at Politecnico di Torino in 2004. In 2007 she concluded her PhD in Industrial Production System Engineering, presenting a thesis on collaboration in complex industrial networks. Now she is working in the Technology to Business Intelligence unit of Istituto Superiore Mario Boella on the theme of information and communication technology diffusion.

Alexandre DOLGUI is Full Professor, Director of the Centre for Industrial Engineering and Computer Science at the École des Mines de Saint-Étienne (France). Faculty Member at the State University of Informatics and Radioelectronics, Belarus (1986–1994) and at the University of Technology of Troyes, France (1995–2003). He has held visiting appointments at the INRIA-Lorraine, project SAGEP, France (from 1992 to 1993 and from 1994 to 1995) and at the Queen's School of Business, Kingston, Canada (in 2005). Research issues on manufacturing line design, production planning, and supply chain optimization. Member of the Board of IFPR, and of IFAC Technical Committees 5.1 and 5.2, Associate Editor or an Editorial Board Member of nine international journals.

Peter P. GROUMPOS received his PhD in 1978 from the State University of New York at Buffalo. He is the President and CEO of the Patras Science Park. Currently he is Professor in the Division of Systems and Control and Director of the Laboratory for Automation and Robotics (LAR-UoP). His main research interests are intelligent manufacturing systems, supervisory hybrid control, soft computing control methods, simulation and application of informatics in a number of areas. He has published over 180 journal and conference papers, book chapters and technical reports.

Geza HAIDEGGER, PhD, Eur-Eng, graduated from the Technical University of Budapest in 1978, and began to work at the Hungarian Academy of Sciences, Computer and Automation Institute. He got his PhD in 2002. He has been in charge of the design of industrial control systems, robots, etc. Member of several scientific societies, author/co-author in more than 150 publications. Project leader of several Hungarian and joint European projects.

Cathal HEAVEY is a Senior Lecturer in the Manufacturing and Operations Engineering Department in the College of Engineering at the University of Limerick. He lectures in the areas of operations research, information technology, supply chain modelling and discrete event simulation. He is Joint Director of the Enterprise Research Centre. His research interests include: simulation modelling of discrete event systems, modelling and analysis of supply chains and manufacturing systems and enterprise modelling, where he has specialized in the area of modelling and optimization of manufacturing and supply chain systems using both analytical and simulation techniques.

Jean-Claude HENNET is a Research Director of the French Research Center (CNRS). He is the head of the research team on Discrete Event Systems and Manufacturing Systems of LSIS in Marseilles, France. He is the author of more than 100 articles in scientific journals and international conferences. His research interests are in the fields of systems theory, optimization, game theory, control under constraints, discrete event decision systems and manufacturing systems.

George L. KOVACS has been a Professor at the Technical University of Budapest since 1995. In 1997 he received the Doctor of the Academy degree from the Hungarian Academy of Sciences (HAS). He has been with the Computer and Automation Research Institute of HAS since 1966, recently head of the CIM Research Laboratory. Author of more than 300 publications, he is a member of several Hungarian and international scientific organizations.

Paul LISTON is a Postdoctoral Researcher at the Enterprise Research Centre at the University of Limerick. He received a PhD, in the area of decision support for contract costing, from the Manufacturing and Operations Engineering Department at the University of Limerick. His research interests include web-based discrete event simulation, supply chain design and analysis, environmental impact performance metrics, decision support and knowledge management systems, non-hierarchical networks, and outsourced services modelling.

Vittorio MARCHIS is a Full Professor of Theoretical and Applied Mechanics (1990) at the Politecnico of Turin, and the Director of the Historical Documentation Centre and Museum of the Politecnico of Turin. Presently he is a Full Professor of History of Technology at the Politecnico di Torino. He organized several scientific conferences and symposia, and he directed many historical exhibitions.

Nicola PASQUINO is a Researcher at the University of Salerno. His current research focuses on production systems.

Thomas POTINECKE studied Mechanical Engineering at the University of Stuttgart. He is a business engineer at the Fraunhofer Institute for Industrial Engineering (IAO) in Stuttgart. He has been working as a scientific researcher in the Competence Centre Innovation Management since 2000. He has gathered expertise in assessment topics concerning virtual engineering, digital manufacturing, process management as well as innovation and networking processes.

Thorsten ROGOWSKI is a business engineer at the Institute of Technology Management (IAT) of the University of Stuttgart. He has worked as a scientific researcher in the Competence Centre Innovation Management since 2002. He has gathered expertise in assessment topics concerning innovation and networking processes as well as the innovation capability of companies. He has coordinated the Collaborative Research Center 374 (Sfb 374) "Development and Assessment of Innovative Products".

Marta SALVADOR received a MSc in Communication Studies at University of Bologna in 2000, and specialized in Communication of Science at the International School for Advanced Studies (SISSA-ISAS) in Trieste in 2003. Knowledge Manager and Communication Expert at EIDON S.p. a. from 2000 to 2008, she has been working on software development and multimedia design of e-learning applications, RTD and innovation project management, as well as in science communication activities. In the field of EU-funded RTD projects, her professional interests are mainly focused on coordination, dissemination and training initiatives.

Stefano SALVADOR received a MSc degree in Contemporary Philosophy at Ca' Foscari University of Venice in 2004; after that he worked in the ICT Sector in Italy and Germany. At the Academia-Industry Interaction Responsible at EIDON S.p. a. from 2006 to 2008, he has also carried out research work for the University of Trieste. Member of the Scientific Committee of the Marie Curie supported CREATE project and coordinator of the first CREATE Training Course, he is currently working in the HR field. Among his professional interests are innovative organizational practices and knowledge management.

Chrysostomos STYLIOS holds a Diploma in Electrical Engineering from the Aristotle University of Thessaloniki (1992) and a PhD degree from University of Patras, Greece (1999). He is a consultant at Patras Science Park and an Assistant Professor at the Department of Informatics and Telecommunication Technology, TEI of Epirus and Director of Knowledge and Intelligent Computing Laboratory. His research interests include soft computing, computational intelligence, modelling complex systems, intelligent systems, decision support systems and artificial intelligence techniques for medical applications.

Teresa TAURINO obtained a Master's degree in Mathematical Engineering at Politecnico di Torino in 2005. Since May 2006 she has collaborated with the European Coordination Action CODESNET (COllaborative DEmand and Supply NETworks). Since January 2007 she has been a PhD student in Production Systems Engineering at Politecnico di Torino, Department of Production Systems and Business Economics. Her main research topic is the development of mathematical tools for modelling and analysis of industrial networks performance.

Agostino VILLA is Full Professor of Performance Analysis of Industrial Production Systems, and of Production Planning and Control since 1990 at Politecnico di Torino, Chairman of the Committee for University Performance Evaluation. His research area is in modelling large-scale distributed industrial systems, designing production planning and control architectures, methods and tools for managing industrial innovation processes. Past President and Fellow of the International Foundation for Production Research (IFPR), he is founder and member of the IFAC TC MIM, the IFIP WG 5.7. Regional editor of *Robotics and Computer Integrated Manufacturing* (Elsevier), and member of the Editorial Committees of some other international journals. Coordinator of the CODESNET Project.

Index

T

V

MIX
Papier aus verantwortungsvollen Quellen
Paper from responsible sources
FSC® C105338

FSC
www.fsc.org

If you have any concerns about our products,
you can contact us on
ProductSafety@springernature.com

In case Publisher is established outside the EU,
the EU authorized representative is:
Springer Nature Customer Service Center GmbH
Europaplatz 3, 69115 Heidelberg, Germany

Printed by Libri Plureos GmbH
in Hamburg, Germany